Signal Conditioning and PC-based Data Acquisition Handbook

the smart approach to instrumentation™

Published 1997
Printed in the United States of America

ISBN (Paperback): 0-9656789-0-3

Signal Conditioning and PC-based Data Acquisition Handbook

Preface

Any industry that performs testing or monitoring has an array of transducers specifically designed for its particular measurement requirements. The intent of this handbook is to introduce the reader to the most commonly used transducer interfaces and to provide practical information for dealing with the most frequently encountered transducers and their associated signals.

Transducers are available for measuring many physical quantities, such as temperature, pressure, strain, vibration, sound, humidity, flow, level, velocity, charge, pH, and chemical composition, among others. In most cases, the transducer manufacturer provides application notes on the transducer's use and principles of operation. The main questions to consider when selecting a transducer are:

- What are the electrical characteristics (amplitude, frequency, source impedance) of the transducer's output?
- What kind of power supply/excitation is required?
- What is the transducer's specified accuracy?
- Over what range of amplitude and frequency are the measurements accurate?
- How is the transducer calibrated?
- How can transducer accuracy and calibration be verified?
- In what environment (temperature, humidity, vibration, pressure) is the transducer able to operate?

It is generally unwise to view a transducer as a black box that provides a specified output for a certain input. Knowing how a transducer works is imperative to making reliable measurements.

Chapter 1
Introduction

This handbook is intended for newcomers to the field of data acquisition and signal conditioning. Emphasis is given to general discussions of ADC measurement and the signal conditioning requirements of selected transducer types. For more detailed descriptions of the signal conditioning schemes discussed, contact IOtech for Applications Notes and Product Specification Sheets. Additional information on IOtech products is available in Chapter 9, a product selection guide.

Most measurements begin with a transducer, a device that converts a measurable physical quality, such as temperature, strain, or acceleration, to an electrical signal. Transducers are available for a wide range of measurements, and come in a variety of shapes, sizes, and specifications. This book is intended to serve as a primer for making measurements by interfacing transducers to a computer using signal conditioning.

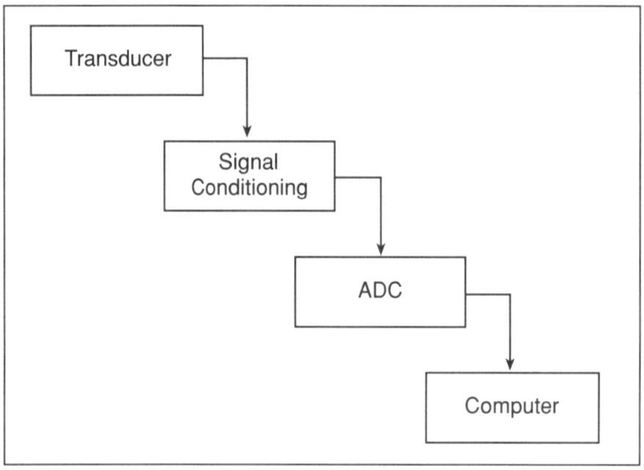

Fig. 1.01: Generic signal-conditioning scheme

Signal conditioning converts a transducer's signal so that an analog-to-digital converter (ADC) can measure the signal. Signal conditioning can include amplification, filtering, differential applications, isolation, simultaneous sample and hold (SS&H), current-to-voltage conversion, voltage-to-frequency conversion, linearization and more. Signal conditioning also includes excitation or bias for transducers that require it.

Figure 1.01 depicts a generic data acquisition signal-conditioning configuration. The transducer is connected to the input of the signal conditioning electronics. The output of the signal conditioning is connected to an ADC input. The ADC converts the analog voltage to a digital signal, which is transferred to the computer for processing, graphing, and storage.

Analog-to-Digital Conversion. Chapter 2 includes a discussion of the four basic ADC types, as well as issues such as accuracy, noise reduction, and discrete sampling considerations. Topics such as input and source impedance, differential voltage measurements, simultaneous sample and hold, selectable input ranges, multiplexing, and isolation are also discussed. A section on discrete sampling considerations, which covers aliasing, windowing, fast Fourier transforms (FFTs), standard Fourier transforms, and digital filtering, is also included.

Multiplexing. Chapter 3 includes a discussion on multiplexing, current measurements, and the associated issues such as simultaneous sample and hold, input buffering, and methods of range and gain selection.

Signal Conditioning by Chapter. Signal conditioning for a wide variety of transducer types, including temperature, strain, force, torque, pressure, and acceleration, is discussed in Chapter 4 through Chapter 8. The temperature measurement section in Chapter 4 describes the principles of operation, signal conditioning, linearization, and accuracy of thermocouples, RTDs, and integrated circuits (ICs).

The strain gage section in Chapter 5 discusses the Wheatstone bridge, as well as the use of strain gages in quarter-bridge, half-bridge, and full-bridge configurations. The use of strain gages in load cells is described, along with excitation and signal conditioning requirements. The piezoelectric transducers (PZTs) section in Chapter 5 describes the use of these devices in voltage and charge amplification configurations and with low- and high-impedance transducers. The pressure transducer section covers both strain-diaphragm transducers used for quasi-static pressure measurements and PZT-based pressure transducers used in dynamic measurements.

Chapter 6 begins with a brief review of general amplification. This section then describes data acquisition front ends, source impedance and multiplexing, filters, and single-ended and differential measurements. The measurement of high voltages and DC and AC currents is also discussed.

Chapter 7 discusses noise reduction and isolation, including specific methods of isolation, such as magnetic, optical, and capacitive.

This handbook also describes digital signal conditioning in Chapter 8, including speed and timing issues.

Frequency measurements, pulse counting, and pulse timing are covered in Chapter 8. The frequency of a signal can be measured using two methods: conversion to a voltage that is read by an ADC, or gated pulse counting. Both methods are described, as is the use of counters for timing applications.

Chapter 2
Analog-to-Digital Conversion

This chapter examines general considerations for analog-to-digital converter (ADC) measurements. Discussed are the four basic ADC types, providing a general description of each while comparing their speed and resolution. Issues such as calibration, linearity, missing codes, and noise are discussed, as are their effects on ADC accuracy.

This chapter also includes information on simultaneous sample and hold (SS&H) and selectable input ranges. Finally, this chapter contains a section on discrete sampling, which includes Fourier theory, aliasing, windowing, fast Fourier transforms (FFTs), standard Fourier transforms, and digital filtering.

ADC Types

An ADC converts an analog voltage to a digital number. The digital number represents the input voltage in discrete steps with finite resolution. ADC resolution is determined by the number of bits that represent the digital number. An n-bit ADC has a resolution of 1 part in 2^n. For example, a 12-bit ADC has a resolution of 1 part in 4096 (2^{12}=4,096). Twelve-bit ADC resolution corresponds to 2.44 mV for a 10V range. Similarly, a 16-bit ADC's resolution is 1 part in 65,536 (2^{16}=65,536), which corresponds to 0.153 mV for a 10V range.

Many different types of analog-to-digital converters are available. Differing ADC types offer varying resolution, accuracy, and speed specifications. The most popular ADC types are the parallel (flash) converter, the successive approximation ADC, the voltage-to-frequency ADC, and the integrating ADC. Descriptions of each follow.

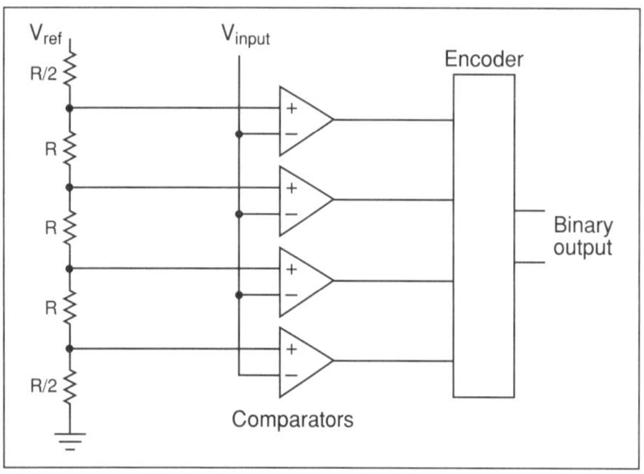

Fig. 2.01: 2-bit parallel converter

Parallel (Flash) Converter

The parallel converter is the simplest ADC implementation. It uses a reference voltage at the full scale of the input range and a voltage divider composed of $2^n + 1$ resistors in series, where n is the ADC resolution in bits. The value of the input voltage is determined by using a comparator at each of the 2^n reference voltages created in the voltage divider. Figure 2.01 depicts a 2-bit parallel converter.

Flash converters are very fast (up to 500 MHz) because the bits are determined in parallel. This method requires a large number of comparators, thereby limiting the resolution of most parallel converters to 8 bits (256 comparators). Flash converters are commonly found in transient digitizers and digital oscilloscopes.

Successive Approximation ADC

A successive approximation ADC employs a digital-to-analog converter (DAC) and a single comparator. It effectively makes a bisection or binomial search by beginning with an output of zero. It provisionally sets each bit of the DAC, beginning with the most significant bit. The search compares the output of the DAC to the voltage being measured. If setting a bit to one causes the DAC output to rise above the input voltage, that bit is set to zero. A diagram of a successive approximation ADC is shown in Figure 2.02.

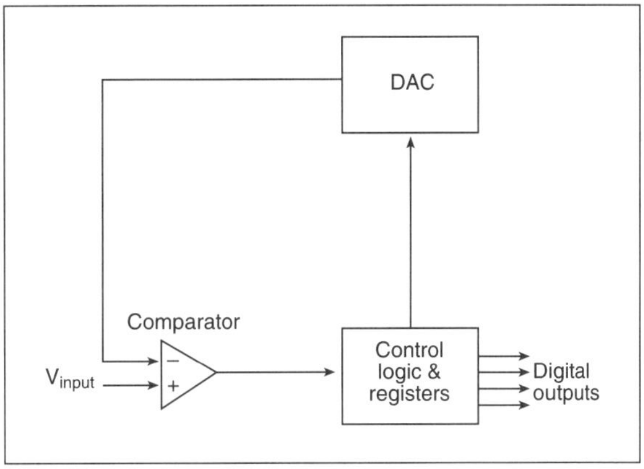

Fig. 2.02: Successive approximation ADC

Successive approximation is slower than flash conversion because the comparisons must be performed in a series, and the ADC must pause at each step to set the DAC and wait for it to settle. However, conversion rates over 200 kHz are common. Successive approximation is relatively inexpensive to implement for 12- and 16-bit resolution. Consequently, they are the most commonly used ADCs, and can be found in many PC-based data acquisition products.

Voltage-to-Frequency ADC

Figure 2.03 depicts the voltage-to-frequency technique. Voltage-to-frequency ADCs convert an input voltage to an output pulse train with a frequency proportional to the input voltage. Output frequency is determined by counting pulses over a fixed time interval, and the voltage is inferred from the known relationship.

Voltage-to-frequency conversion has a high degree of noise rejection, because the input signal is effectively integrated over the counting interval. Voltage-to-frequency conversion is commonly used to convert slow and often noisy signals.

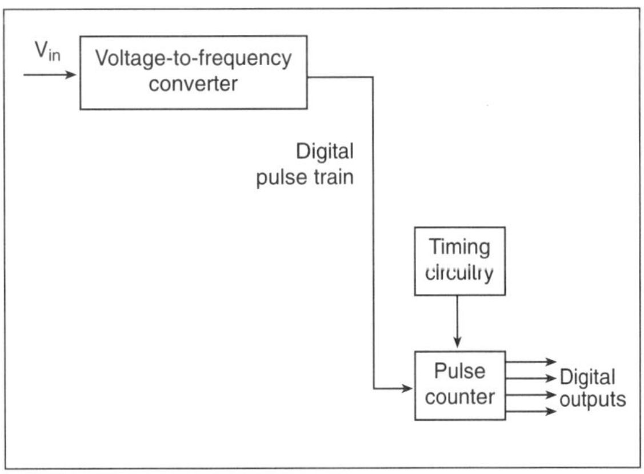

Fig. 2.03: Voltage-to-frequency ADC

It is also useful for remote sensing applications in noisy environments. The input voltage is converted to a frequency at the remote location, and the digital pulse train is transmitted over a pair of wires to the counter. This eliminates the noise that can be introduced in the transmission of an analog signal over a long distance.

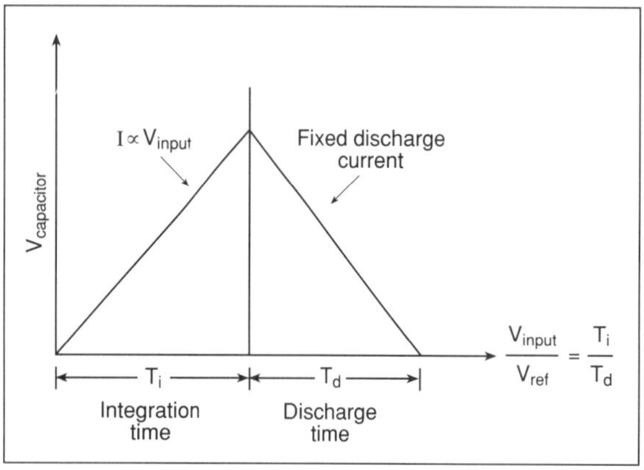

Fig. 2.04: Integration and discharge of an integrating ADC

Integrating ADC

A number of ADCs use integrating techniques, which measure the time to charge or discharge a capacitor to determine input voltage. Figure 2.04 shows "Dual-slope" integration, a common integration technique. Using a current that is proportional to the input voltage, a capacitor is charged for a fixed time period. The average input voltage is determined by measuring the time required to discharge the capacitor using a constant current.

Integrating the ADC input over an interval reduces the effect of noise pickup at the AC line frequency if the integration time is matched to a multiple of the AC period. For this reason, it is commonly used in precision digital multimeters and panel meters. Twenty-bit accuracy is not uncommon. The disadvantage is a relatively slow conversion rate (60 Hz maximum, slower for ADCs that integrate over multiple line cycles).

Summary of ADC Types

Figure 2.05 summarizes the previously discussed ADC types and their resolution and speed ranges.

ADC Type Table		
ADC Type	**Typical Resolution**	**Typical Speed**
Parallel Converter	4-8 bit	100 kHz-500 MHz
Successive Approximation	8-16 bit	10 kHz-1 MHz
Voltage-to-Frequency	8-12 bit	1-60Hz*
Integrating	12-24 bit	1-60 Hz*

* With line cycle rejection

Fig. 2.05: Summary of ADC types

Accuracy

Accuracy is an important consideration when selecting an ADC for use in test and measurement applications. The following section provides an in-depth discussion of accuracy considerations, pertaining to resolution, calibration, linearity, missing codes, and noise.

Accuracy vs. Resolution

The accuracy of a measurement is influenced by a variety of factors. If each independent error is σ_i, the total error is

$$\sigma_{total} = \sqrt{\Sigma_i \sigma_i^2}$$

This calculation includes errors resulting from the transducer, noise pickup, ADC quantization, gain, offset, and other factors.

ADC resolution error is termed quantization error. In an ideal ADC, any voltage in the range that corresponds to a unique digital code is represented by that code. The error in this case is half of the least significant bit (LSB) at most. For a 12-bit ADC with a 10V range, this error is 2.44 mV (0.0244%). There are three common methods of specifying a contribution to ADC error: the error in least significant bits (LSBs), the voltage error for a specified range, and the percent-of-reading error. It is important to recognize that most ADCs are not as accurate as their specified resolution, because quantization error is only one potential source of error. Nonetheless, the accuracy of a good ADC should approach its specified resolution.

For more information concerning accuracy, refer to the Calibration section, which follows. For an in-depth discussion of errors arising in particular transducers, refer to Chapters 3 through 8.

Figure 2.06 illustrates common error types encountered when using a 3-bit ADC. If the manufacturer provides calibration procedures, offset and gain errors can usually be reduced to negligible levels, as discussed below. However, errors in linearity and missing codes will contribute to the overall error.

Calibration

There are several common methods for calibrating an ADC. In hardware calibration, the offset and gain of the instrumentation amplifier that serves as the ADC front end is adjusted with trim pots. (The gain of the ADC can also be adjusted by changing the reference voltage.) In hardware/software calibration, digital-to-analog converters that null the offset and set the full scale voltages are programmed via software. In software calibration, there is no hardware adjustment. Calibration correction factors are stored in the nonvolatile memory of the data acquisition system or in the computer and used to convert the reading from the ADC.

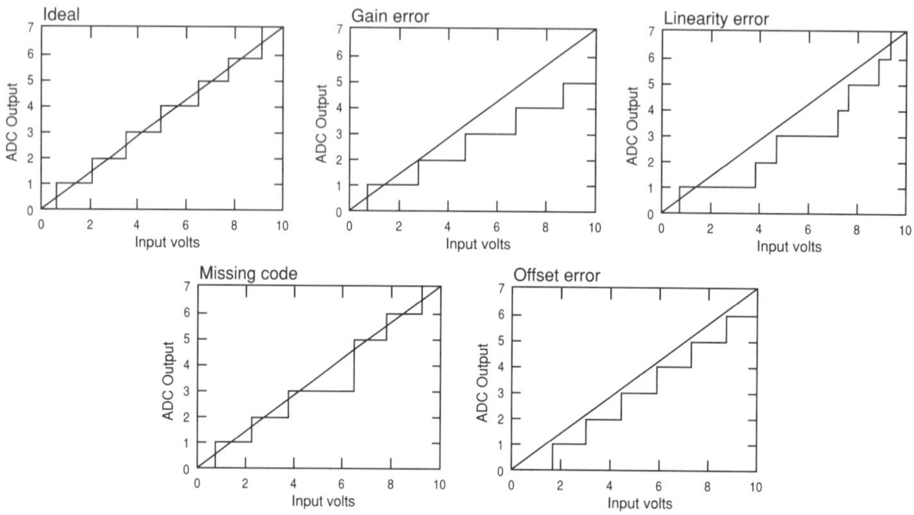

Fig. 2.06: Common ADC error sources. The straight line is ideal output from an ADC with infinite-bit resolution. The step function shows the indicated error for a 3-bit ADC.

Even if an ADC is calibrated at the factory, it will need to be calibrated again after a period of time (typically six months to a year, but possibly more frequently for ADCs of greater resolution than 16 bits). Variations in the operating temperature can also affect instrument calibration. Calibration procedures vary but usually require either a known reference source or a meter of greater accuracy than the device being calibrated. Typically, offset is set via a 0V input, and gain is set via a full-scale input.

In many measurements, the voltage is not the physical quantity under test. Consequently, it may be preferable to calibrate the complete measurement system rather than its individual parts. For example, consider a load cell for which the manufacturer specifies the output for a given load and excitation voltage. One could calibrate the ADC and combine this with the manufacturer's specification and a measurement of the excitation voltage; however, this technique is open to error. Specifically, three distinct error sources are possible in this technique: error in the ADC calibration, error in the manufacturer's specifications, and error in the measurement of the excitation voltage. To circumvent these error sources, one can calibrate the measurement system using known loads and obtain a direct relationship between load and ADC output.

Linearity

If input voltage and ADC output deviate from the diagonal lines in Figure 2.06 more than the ideal step function, the result is ADC error that is nearly impossible to eliminate by calibration. This type of ADC error is referred to as nonlinearity error. If nonlinearity is present in a calibrated ADC, the error is often largest near the middle of the input range, as shown in Figure 2.06. The nonlinearity in a good ADC should be 1 LSB or less.

Missing Codes

Some ADCs have missing codes. In Figure 2.06, the ADC does not provide an output of four for any input voltage. This error can result in a significant loss in resolution and accuracy. A quality ADC should have no missing codes.

Noise

Many users are surprised by noise encountered when measuring millivolt signals or attempting accurate measurements on larger signals. Investing in an accurate ADC is only the first step in accurately measuring analog input signals. Controlling noise is imperative.

Many ADCs reside on cards that plug into a PC expansion bus, where electrical noise can present serious problems. Expansion bus noise often far exceeds the ADC's sensitivity resulting in significant loss of measurement accuracy. Placing the ADC outside the PC is often a better solution. An ADC in an external enclosure can communicate with the computer over an IEEE 488 bus, serial port, or parallel port. If an application requires placement of the ADC within the computer, the noise level should be tested by connecting the ADC input to signal common and observing deviations in ADC output. (Connecting the ADC input to signal common isolates the cause of the noise to the circuit card. More careful diagnostics are necessary when using an external voltage source because noise can arise from the external source and from the input leads.)

Noise Reduction and Measurement Accuracy

One technique for reducing noise and ensuring measurement accuracy is with isolation, which also eliminates ground loops. Ground loops occur when two or more devices in a system, such as a measurement instrument and a transducer, are connected to ground at different physical locations. Slight differences in the actual potential of each ground results in a current flow from one device to the other. This current, which often flows through the low lead of a pair of measurement wires, generates a voltage drop which can directly lead to measurement inaccuracies and noise. If at least one device is isolated, such as the measurement device, then there is no path for the current flow, and thereby no contribution to noise or inaccuracy.

Protection

Many data acquisition systems utilize solid-state multiplexing circuitry in order to very quickly scan multiple input channels. These solid-state multiplexers are among the most susceptible circuitry to overload voltages, which can commonly occur in a system. Typically, multiplexers can only accept up to 20 or 30 volts before damage occurs. Other solid-state devices in a measurement system include input amplifiers and bias sources, both of which are also susceptible to damage from over voltage. Isolation is one of a number of techniques used to protect sensitive solid state circuitry in a data acquisition system.

Although isolation does not protect against excessive normal-mode input voltage (voltage across a pair of inputs), it does protect against excessive common-mode voltage. It accomplishes this by eliminating the potentially large current that would otherwise flow from the signal source to the data acquisition system, as a result of the common mode voltage. By eliminating this current flow, the possibility of damage is eliminated.

High common-mode voltage measurements

It is often necessary to measure a small voltage which is residing on another, much larger voltage. For example, if a thermocouple is mounted to one terminal of a battery, then the measurement device must be capable of measuring the microvolt output of the thermocouple while rejecting the battery voltage. If the common-mode voltage is less than 10-15V, a differential measurement via an instrumentation amplifier will read the thermocouple voltage while ignoring the battery voltage. If the common-mode voltage is higher than 10-15V, an isolation method is generally required.

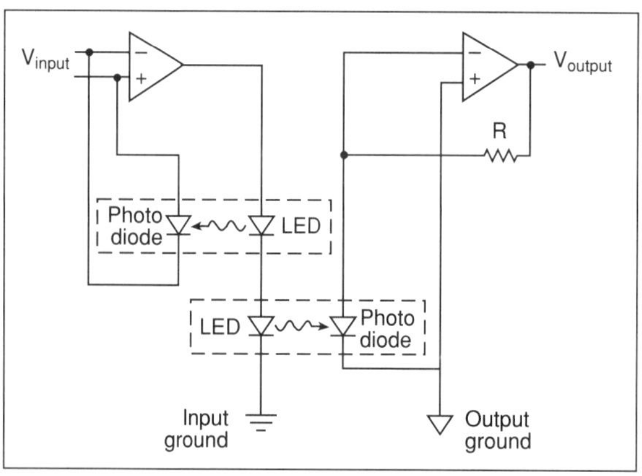

Fig. 2.07: Optically-coupled isolation amplifier

There are several isolation methods with the common characteristic of a high common-mode voltage from input to output. Each channel can have an isolation amplifier, or, a group of channels that are not isolated from each other can be multiplexed and digitized by an analog-to-digital converter before the digital data is isolated from the remainder of the system.

Actual isolation barriers can be optical, magnetic, or capacitive. The most common are optical schemes in which infrared emission from an LED is detected by a photodiode on the opposite side of a quartz barrier. Figure 2.07 illustrates an optically-coupled isolation amplifier. Optocouplers can be used to transmit pulse trains in which frequency or pulse width vary with analog signal magnitude or which contain numerical data in serial pulse trains. It is even possible to transmit analog information by varying LED current as in Figure 2.07. Magnetic barriers using transformers and capacitive barriers are generally internal and used in monolithic or hybrid isolation amplifiers.

Frequency Coupled Isolation. In frequency coupled isolation, a high frequency carrier signal is inductively or capacitively coupled across the isolation barrier. The signal is modulated on the input side and demodulated on the output side to reproduce the original input signal.

Isolated ADC. When using an isolated ADC, the ADC and accompanying signal conditioning are floated. The input signal is converted to a digital signal by the ADC and the interface for transferring the digital code is digitally isolated. See Chapter 6 for a detailed discussion.

Discrete Sampling Considerations

The Nyquist sampling theorem says that if a signal only contains frequencies less than cutoff frequency f_c, all the information in the signal can be captured by sampling at $2f_c$. The upshot of this is that capturing a signal with maximum frequency component f_{max} requires sampling at a rate of at least $2f_{max}$. In practice, it is best to set the sampling rate between three and five times the signal's highest frequency component. One reason is to retain accuracy at the signal's higher frequency components.

Aliasing

Aliasing can also be seen in the time domain. Figure 2.09 shows a 1-kHz sine wave sampled at 800 Hz. The apparent frequency of the sine wave is much too low. Figure 2.08 shows the result of sampling the same 1-kHz sine wave at 5 kHz. The sampled wave appears to have the correct frequency.

Aliasing is the main reason to sample at a rate higher than the Nyquist frequency. Aliasing—the generation of false, low-frequency signals—occurs when an ADC's sampling rate is too low. Input signals are seldom bandwidth limited with zero amplitude higher than f_{max}. A signal with frequency components higher than one-half the sampling frequency will cause the amplitude to appear below one-half the sampling frequency in the Fourier transform. This is called aliasing, and it can cause inaccuracies in sampled signal and also in the Fourier transform.

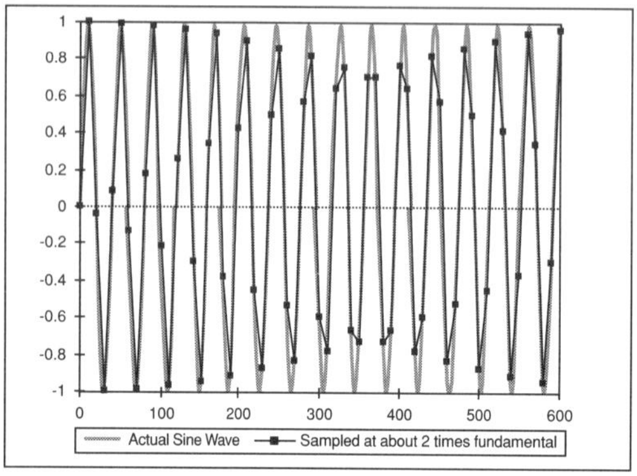

Fig. 2.08: When sampled at just over 2 times the frequency of the sine wave, the acquired waveform closely approximates the actual signal

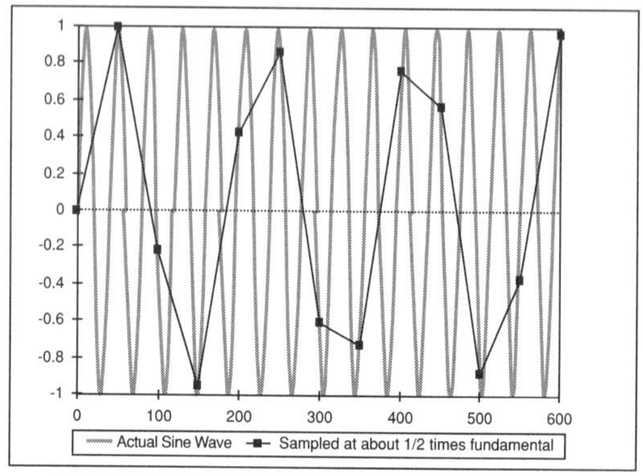

Fig. 2.09: When sampling too slowly, the acquired waveform erroneously represents the real sine wave

Aliasing is illustrated in Figure 2.10, which shows a square wave's Fourier transform. For the purpose of illustration, assume that the experiment has been designed to provide only frequencies of under 2 kHz. Ideally, a Fourier transform of a 500-Hz square wave contains one peak at 500 Hz, and another at 1500 Hz, which is one third the height of the first peak. In Figure 2.10, however, higher frequency peaks are aliased into the Fourier transform's low-frequency range.

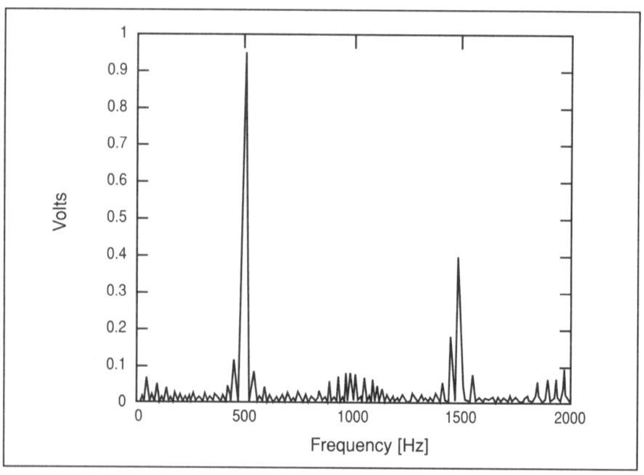

Fig. 2.10: Fourier transform of a 500-Hz square wave sampled at 4 kHz with no filtering

The use of a low-pass filter at 2 kHz, as shown in Figure 2.11, removes most of the aliased peaks. Low-pass filters used for this purpose are often called "anti-aliasing" filters.

When the sampling rate is increased to four times the highest frequency of interest, the Fourier transform in the range of interest looks even better. Although a small peak remains at 1,000 Hz, it is probably the result of an imperfect square wave rather than an effect of aliasing. See Figure 2.12.

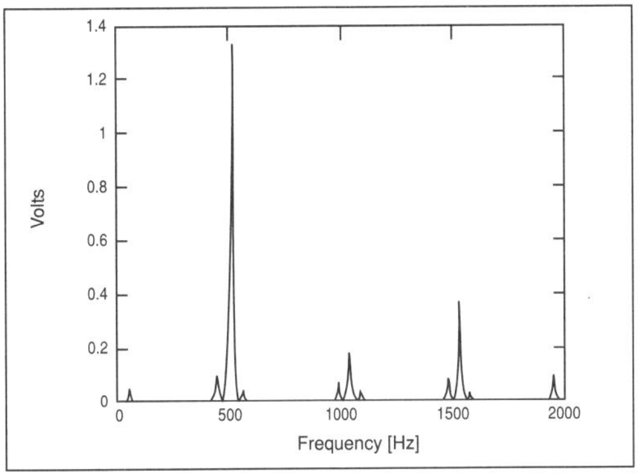

Fig. 2.11: Fourier transform of 500-Hz square wave sampled at a 4 kHz with low-pass filter cutoff at 2 kHz

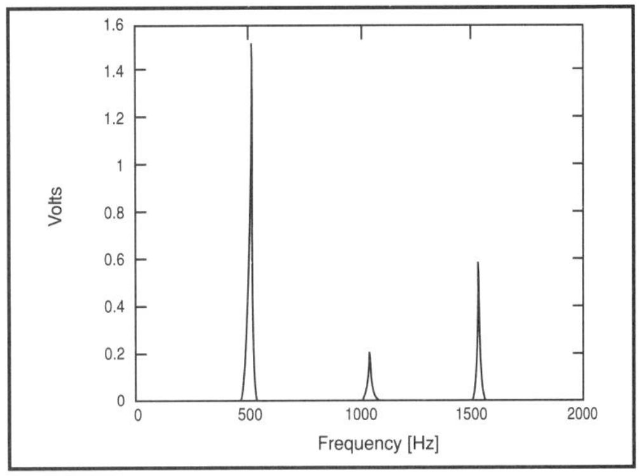

Fig. 2.12: Fourier transform of 500-Hz square wave sampled at 8 kHz with low-pass filter cutoff at 2 kHz

Windowing

Windowing is the multiplication of the input signal with a weighting function to reduce spurious oscillations in a Fourier transform. Real measurements are performed over finite time intervals. In contrast, Fourier transforms are defined over infinite time intervals. As such, the Fourier transform of sampled data is an approximation. Consequently, the resolution of the Fourier transform is limited to roughly 1/T, where T is the finite time interval over which the measurement was made. Fourier transform resolution can only be improved by sampling for a longer interval.

Using a finite time interval also causes spurious oscillations in the Fourier transform. From a mathematical viewpoint, spurious oscillations are caused by the signal being instantaneously turned on at the beginning of the measurement and then suddenly turned off at the end of the measurement. Figure 2.13 illustrates an example of spurious oscillations.

Implementing window functions can help minimize spurious oscillations of a signal. Multiplying the sampled data by a window function that rises gradually from zero decreases the spurious oscillations at the expense of a slight loss in triggering resolution. There are many possible window functions, all of which involve trade-offs between amplitude estimation and frequency resolution.

Fast Fourier Transforms

The fast Fourier transform (FFT) is so common today that "FFT" has become an imprecise synonym for Fourier transforms in general. The FFT is a digital algorithm for com-

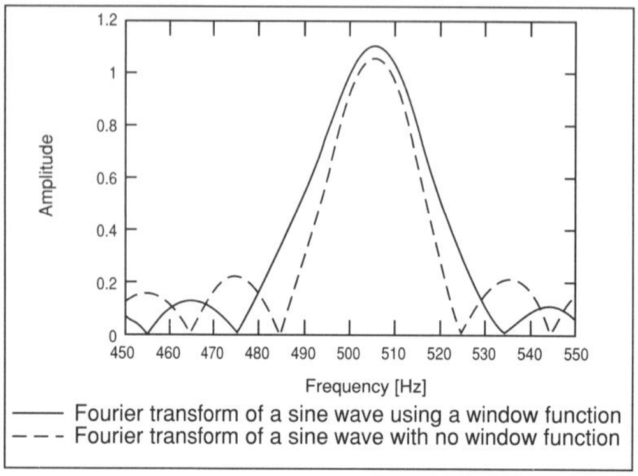

Fig. 2.13: A Fourier transform with window function and without window function

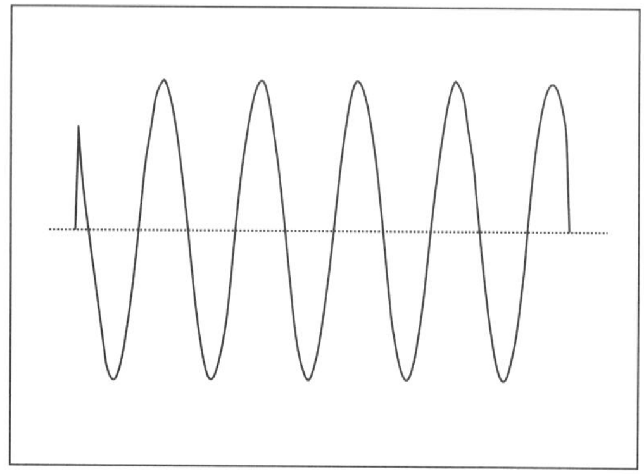

Fig. 2.14: Without window function, abrupt beginning and end points of waveform produce erroneous frequency information

puting Fourier transforms of data discretely sampled at a constant interval. The FFT's simplest implementation requires 2^n samples. Other implementations accept other special numbers of samples. If the data set to be transformed has a different number of samples than required by the FFT algorithm, the data is often padded with zeros to achieve the required number. This leads to inaccuracies, but they are often tolerable.

Standard Fourier Transforms

A standard Fourier transform (SFT) can be used in applications where the number of samples cannot be arranged to fall on one of the special numbers required by an FFT. The SFT is also useful for applications that cannot tolerate the inaccuracies introduced by padding with zeros, a handicap of the FFT.

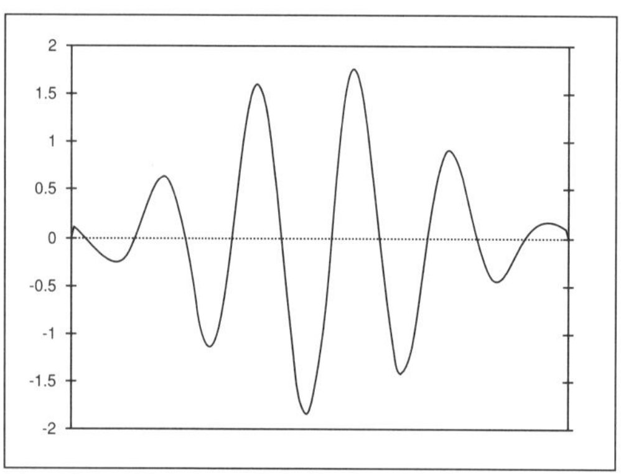

Fig. 2.15: When multiplied by a window function, the frequency information is preserved while minimizing the affect of irregularities at the beginning and end of the sample segment

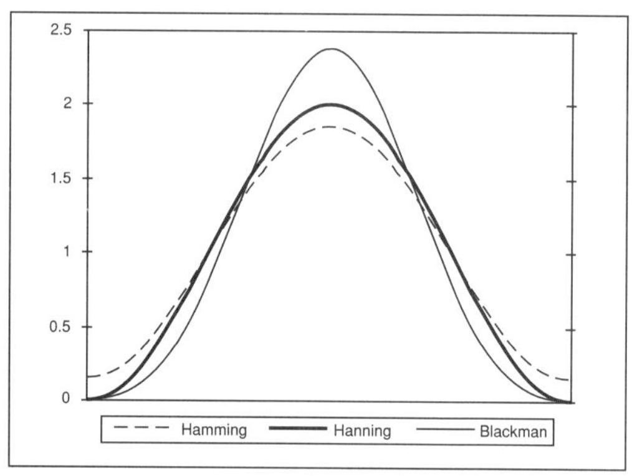

Fig. 2.16: Three common window types

The SFT is also suited to applications where the data is not sampled at evenly spaced intervals or where sample points are missing. Finally, the SFT can be used to provide more closely spaced points in the frequency domain than can be obtained with an FFT. (In an FFT, adjacent points are separated by half the sampling frequency. Points arbitrarily close in frequency can be obtained using an SFT.)

There are many standard numerical integration techniques available for computing SFTs from sampled data. Whichever technique is selected for the problem at hand, it will probably be much slower than an FFT of a similar number of points. This is becoming less of an issue, however, as the speed of modern computers increases.

Digital Filtering

Digital filtering is accomplished in three steps. First, the signal must be subjected to a Fourier transform. Then, the signal's amplitude in the frequency domain must be multiplied by the desired frequency response. Finally, the transferred signal must be inverse Fourier transformed back into the time domain. Figure 2.17 shows the effect of digital filtering on the noisy signal. Note that the solid line represents the unfiltered signal, while the two dashed lines represent different digital filters.

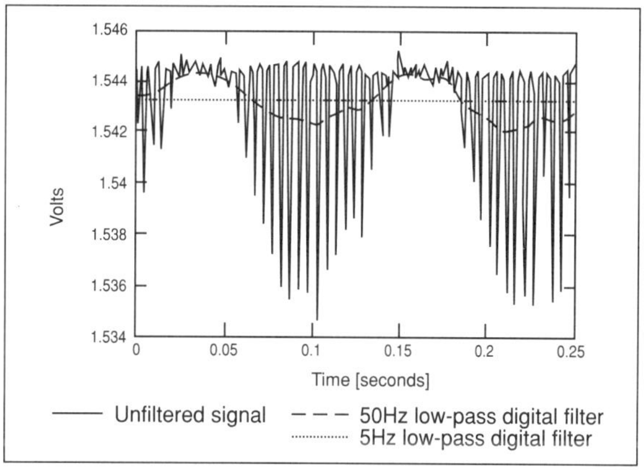

Fig. 2.17: The effect of digital filtering on the noisy signal

Digital filtering is advantageous because the filter itself can be tailored to any frequency response without introducing phase error. In contrast, analog filters are only available with a few frequency response curves, and all introduce some element of phase error. Digital filtering has one disadvantage, however; it cannot be used for anti-aliasing. Only analog filters may be used for anti-aliasing.

Sampling Hints

There are several important "sampling hints" to observe when designing an application. These hints are not absolutes nor do they guarantee optimal results. However, they do provide a useful starting point for planning frequency analysis of a physical process. These sampling hints include:

- A Fourier transform's highest meaningful frequency is one-half of the sampling frequency
- The sampling rate should be at least three to five times the highest frequency of interest
- An anti-aliasing low-pass filter is typically required; the cutoff frequency should be close to the signal's highest frequency of interest
- Digital filtering can be used to smooth the data or to remove noise in a specified range after acquisition; however, aliasing can only be prevented with an analog low-pass filter
- If the phase relationship between multiple signals is important, a simultaneous sample and hold circuit should be used (see multiplexing in Chapter 3)
- Fourier transform resolution is inversely proportional to measurement time; acquiring data over a longer period of time results in narrower peaks in the Fourier transform.

Chapter 3
Multiplexing

Multiplexing

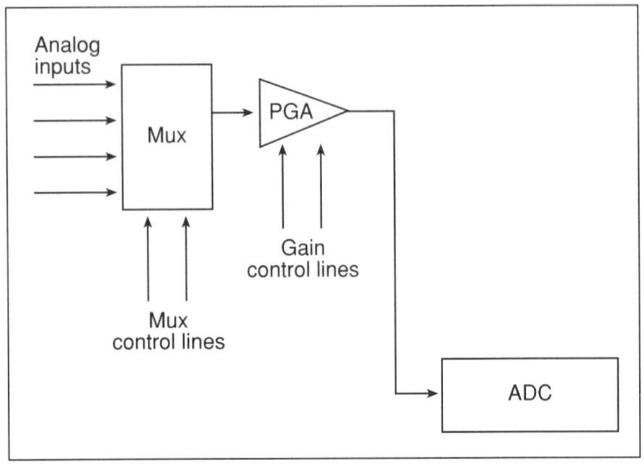

Fig. 3.01: The multiplexer (Mux) is a fast switch that directs different input signals to the ADC for digitizing

A multiplexer is a switch that allows a single ADC to measure many input channels. This switch can be implemented via relays or solid-state switches. A relay is a mechanical switch, so rates are relatively slow (less than 1 kHz for reed relays, which are the fastest type), but large voltages and high isolation (several kV) can be achieved. The current capacity of a relay is determined by its size and contact type. Currents of 3A are typical in relays used in laboratory instruments. Much larger currents can be switched with the larger relays common in industrial applications.

Solid state switches are much faster than relays, and speeds of several MHz are common. However, these small devices are easily destroyed by voltages larger than 25V, and they are a poor choice for isolated applications. Solid-state switches typically support currents less than 1 mA.

A multiplexing scheme such as that shown in Figure 3.01 eliminates the high cost of having multiple ADCs. A programmable gain amplifier (PGA) allows each channel to have a different gain and input range. Multiplexing reduces the rate at which data can be acquired from an individual channel, because multiple channels are scanned sequentially. For example, an ADC that can sample a single channel at 100 kHz is limited to a 12.5-kHz per-channel sampling rate when sampling eight channels.

IOtech's 100-kHz and 1-MHz data acquisition systems are examples of systems that use software selectable channel and gain sequencing. IOtech's 100-kHz systems provide a 512-location scan sequencer that allows you to select, via software, each channel and its associated input amplifier gain for both the built-in channels and the expansion channels. The sequencer circuitry circumvents a major limitation encountered with many plug-in data acquisition boards—a drastic reduction in the scan rate for expansion channels. All channels are scanned, including expansion channels, at 100 kHz (10 µs/channel). Digital inputs can also be scanned using the same scan sequence employed for analog inputs, enabling the time correlation of acquired digital data to acquired analog data. These products permit each scan group, containing up to 512 channel/gain combinations, to be repeated immediately or at programmable intervals of up to 12 hours. Within each scan group, consecutive channels are measured at a fixed 10 µs/channel rate. Figure 3.02 illustrates a 512-location scan sequencer operating in a 100-kHz data acquisition system.

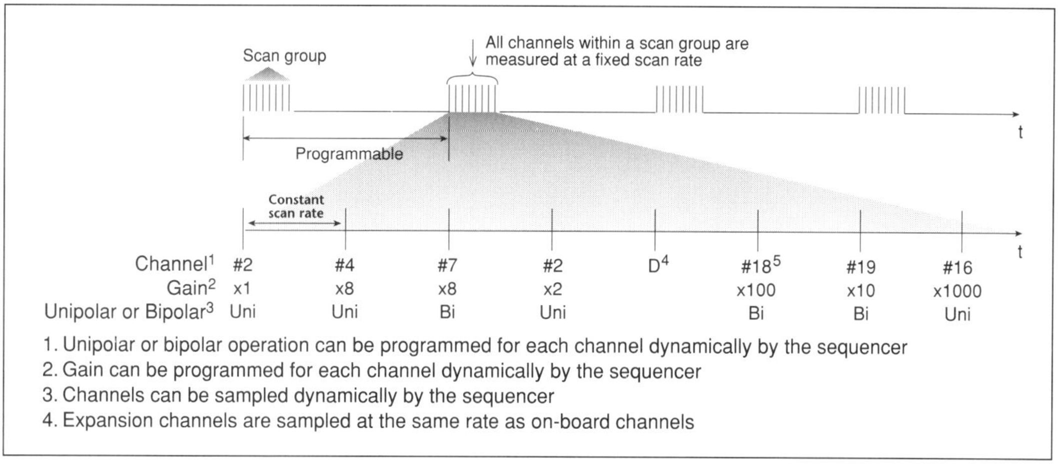

Fig. 3.02: *512-location scan sequencer example*

Unfortunately, multiplexing may introduce problems. A high source impedance can combine with stray capacitance to cause settling problems and crosstalk between channels. Multiplexer impedance can also lead to signal degradation. A solid state multiplexer can have an impedance of tens of Ohms, whereas a relay typically has a resistance of less than 0.01 Ohm.

Sequence vs. Software Selectable Ranges

Most data acquisition system implementations permit different input ranges, although the manner in which they do so varies considerably. Some data acquisition systems allow the input range to be switch or jumper selected on the circuit board. Others provide software selectable gain; this is more convenient, but a distinction should be made between data acquisition systems whose channels must all have the same gain, and systems that allow you to sequentially select the input range of each channel. It is often advantageous to have different input ranges on different channels, especially when measuring signals from different transducers. Thermocouples and strain gages require input ranges of tens of millivolts, while other transducers might output several volts.

A data acquisition system with a software selectable range can be used to measure different ranges on different channels at a relatively slow rate by issuing a software command to change the gain between samples.

There are two problems with this technique. First, it is slow; issuing a software command to change the gain of a PGA can take tens or hundreds of milliseconds, lowering the sample rate to several Hz. Second, the speed of this sequence is often indeterminate, due to variants in the PC instruction cycle times. So cycling through it continuously will give samples with an uneven (and unknown) spacing in time. This complicates time-series analysis and makes FFT analysis impossible since the FFT requires evenly spaced samples.

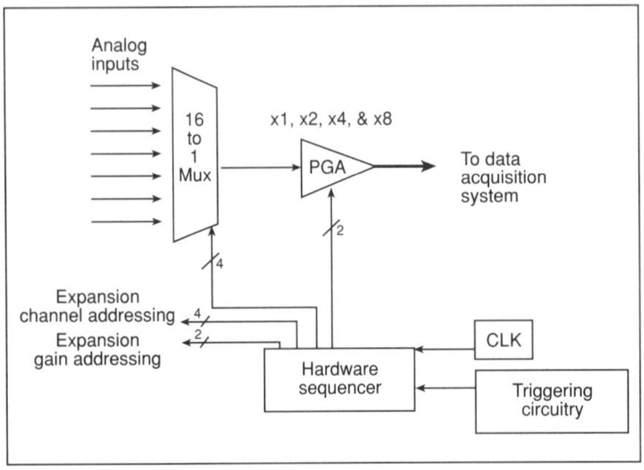

Fig. 3.03: Multiplexer scheme with sequence selectable gain

As discussed earlier, a better implementation provides a sequencer that controls the channel selection *and* gain. The maximum acquisition rate can be achieved with *sequence*-selectable ranges, whereas the acquisition rate will slow considerably with *software*-selectable ranges if channels require different ranges. IOtech's 100-kHz data acquisition systems provide a 512-location scan sequencer that allows you to select each channel and associated input amplifier gain at random. These products permit each scan group to be repeated immediately or at programmable intervals.

Input Buffering

The impact of source impedance and stray capacitance can be estimated via a simple formula: the time constant associated with a source impedance (R) and stray capacitance (C) is T=RC.

For example, suppose you want to know the maximum tolerable input impedance for a 100-kHz multiplexer. The time between measurements on adjacent channels in the scan sequence is 10 µs. In a length of time T=RC, the voltage error decays by a factor of 2.718. Reducing the error to 0.005% requires waiting 10T.

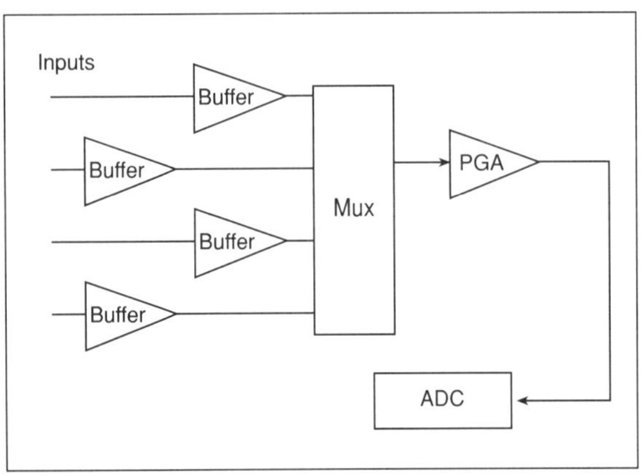

Fig. 3.04: Buffering signals before the multiplexer increases accuracy, especially with high-impedance sources or fast multiplexing

Consequently, a fixed time of 10 µs between scans (T_{scan}) and an error of 0.005% would appear to require a time of T=1 µs. In a typical multiplexed data acquisition system, this would yield errors due to insufficient settling time. The difference can be explained as follows. Most 100-kHz converters require 2 µs (T_{samp}) to sample the input signal. Subtracting this from the scan time results in a settling time of: $T_{settle}=T_{scan}-T_{samp}$, T_{settle}=8 µs.

If we assume a typical 16-bit data acquisition system, the internal settling time (T_{int}) may be 6 μs. The external settling time may then be computed as follows: $T_{ext}=\sqrt{T_{settle}^2-T_{int}^2}$, T_{ext}=5.29 μs.

For a 16-bit data acquisition system with 100 picofarad of input capacitance (C_{in}) and a multiplexer resistance (R_{mux}) of 100 Ohms, the maximum external resistance is calculated as follows: $R_{ext}= (T_{ext}/C_{in}\ 1n(2^{16}))-R_{mux}$, R_{ext}=4,670 Ohms.

The above examples are simplified and do not included any effects due to multiplexer charge injection or inductive reactance in the measurement wiring. In actual practice the practical upper limit on source resistance is between 1,500 and 2,000 Ohms.

Most input signals have source impedances of less than 15 KOhm, so such a maximum source impedance is usually not a problem. However, faster multiplexer rates require lower source impedances. For example, a 1-MHz multiplexer in a 12-bit system requires a source impedance of under 1.5 KOhm. If the source impedance exceeds this value, buffering is necessary to obtain accurate measurements. A buffer is an amplifier with high input impedance and very low output impedance. Placing a buffer on each channel between the transducer and the multiplexer eliminates inaccuracies by preventing the multiplexer's stray capacitance from having to discharge through the impedance of the transducer. IOtech's WaveBook/512 is a high-speed device that uses this configuration. This arrangement is illustrated in Figure 3.04.

Simultaneous Sample and Hold (SS&H)

A multiplexed ADC measurement introduces a time skew among channels, which is intolerable in some applications. Employing simultaneous sample and hold (SS&H) on multiple channels remedies time-skew problems. Simultaneous sample and hold requires that each channel be equipped with a buffer that samples the signal at the beginning of the scan sequence. The signal at the buffer output is held at the sampled value while the multiplexer switches through all channels and the ADC digitizes the readings. In a good simultaneous sample and hold implementation, all channels are sampled within 100 ns of each other.

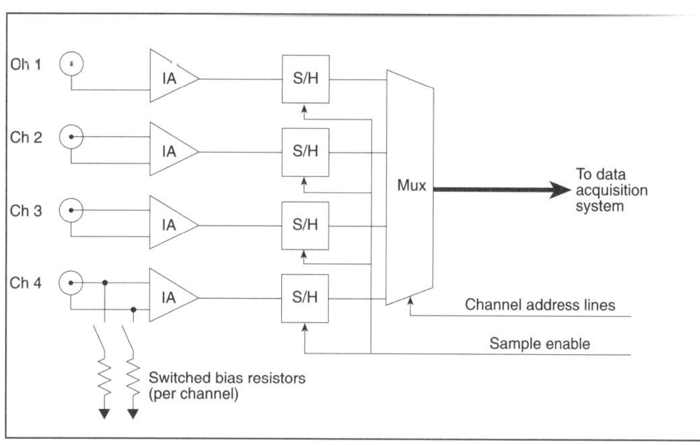

Fig. 3.05: DBK17 simultaneous sample and hold card block diagram

Figure 3.05 shows a common scheme for simultaneous sample and hold; this design is used on IOtech's DBK17 simultaneous sample and hold card, an expansion option for IOtech's 100-kHz data acquisition systems. Each input signal passes through an instrumentation amplifier (IA) and into a sample and hold buffer (S/H). When the sample enable line goes high, each S/H samples its input signal and holds it while the multiplexer switches through the readings. This scheme ensures that all the samples are taken within 50 ns of each other, even with up to sixty-four DBK17s connected to a single instrument. A system configured with sixty-four DBK17s would provide 256 simultaneous channels.

Current Measurements

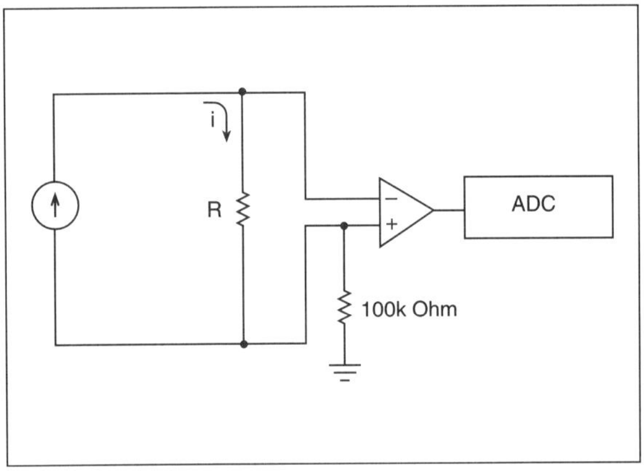

Fig. 3.06: Current measurement using a shunt resistor and a differential amplifier

Many transducers can be configured to output a 4-20 mA current with a linear relationship to the quantity being measured. A shunt resistor converts the current to a voltage for measurement by an ADC. Figure 3.06 illustrates current measurement. If the current (I) is passed through a resistance (R), the voltage across the resistor is V=IR. Consequently, a 500 Ohm resistor is sufficient to map a 20 mA full-scale current into a 10 VFS voltage for measurement with a data acquisition system. Using a transducer configured for current output can yield high noise immunity and accuracy, particularly if there is a large distance between the data acquisition system and the transducer. Long voltage leads pick up noise and suffer an ohmic voltage drop.

The current value inferred from measuring the voltage across a shunt resistor is only as accurate as the resistor. Common resistor accuracies are 5%, 1%, 0.1%, and 0.01%. Resistors are commonly specified according to their accuracy and stability when exposed to changing temperatures. All resistors change with temperature. This phenomenon can be used to measure temperature, as discussed in Chapter 4, however, this effect is undesirable when measuring current. Many resistor manufacturers make an effort to lower the effects of changing temperatures on resistance. This specification means that the actual resistance is within the specified tolerance of the given resistance. In addition, a 0.1% resistor usually has a lower temperature coefficient and better long-term stability than a resistor with lower accuracy, due to its construction technique, which is usually wire wound or metal film vs. carbon composition.

Chapter 4
Temperature Measurement

This chapter examines various transducers for measuring temperature: thermocouples, and RTDs. It also discusses the required signal conditioning, as well as various techniques for optimizing the accuracy of your temperature measurement.

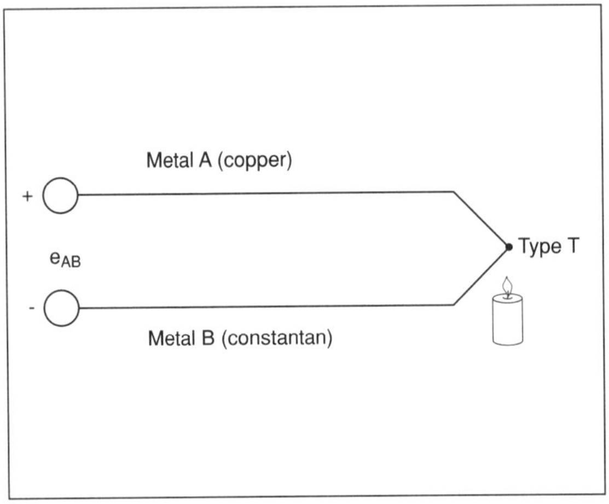

Fig. 4.01: *Illustration of a Type T thermocouple*

Thermocouples

Thermocouples are probably the most widely used and least understood of all temperature measuring devices. Thermocouples provide a simple and efficient means of temperature measurement by generating a voltage that is a function of temperature. All electrically conducting materials produce a thermal electromotive force (emf or voltage difference) as a function of the temperature gradients within the material. This is called the Seebeck effect. The amount of the Seebeck effect depends on the chemical composition of the material used in the thermocouple.

When two different materials are connected to create a TC, a voltage is generated. This voltage is the difference of the voltages generated by the two materials. In principle, a thermocouple can be made from almost any two metals. In practice, several thermocouple types have become de facto standards because they possess desirable qualities, such as highly predictable output voltages and large voltage-to-temperature ratios. Some common thermocouple types are J, K, T, E, N28, N14, S, R, and B. Figure 4.01 depicts a Type T thermocouple. In theory, the temperature can be inferred from such a voltage by consulting standard tables or using linearization algorithms. In practice, this voltage cannot be directly used, because the connection of the thermocouple wires to the measurement device constitutes a thermocouple providing another thermal emf that must be compensated for; cold junction compensation can be used for this purpose.

Cold Junction Compensation. Figure 4.02 shows a thermocouple that has been placed in series with a second thermocouple, which has been placed in an ice bath. This ice bath is called the reference junction. An ice bath can produce an extremely accurate 0°C temperature. The ice bath is normally referred to as the triple point of water and is considered to be a standard temperature for use in thermometry calibration. This reference junction is required because all thermocouple emf tables, as published by NIST

(National Institute of Standards and Technology), are referenced to the emf output of a thermocouple which is held at a temperature of 0.00 degrees Celsius. In figure 4.02, the chromel/copper and alumel/copper thermocouples in the ice bath have a 0.0V contribution to the voltage measured at the meter. The voltage read is entirely from the chromel/alumel thermocouple. The copper wires connected to the copper terminals on the meter do not constitute a thermocouple because they are the same metal connection and both terminals are the same temperature.

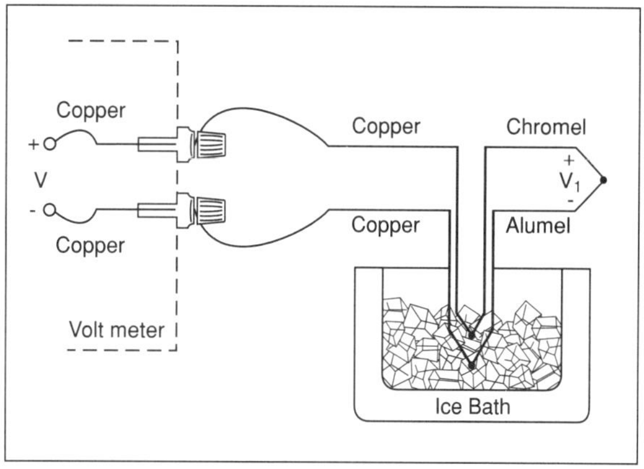

Fig. 4.02: Thermocouple circuit with ice bath

Maintaining an ice bath and an additional reference thermocouple for every thermocouple probe is not practical in most systems. If we know the temperature at the point where we connect to our thermocouple wires, (the reference junction), and if both connecting wires are at the same temperature, then the ice bath can be eliminated. In such a case, the opposing thermoelectric voltage generated by the non-zero temperature of the reference junction can simply be added to the thermocouple emf. This is cold junction compensation (CJC).

CJC is an essential part of accurate thermocouple readings. CJC must be implemented in any system that has no ice-point reference junction. The technique works best if the CJC device is close to the terminal blocks that connect the external thermocouples, and if there are no temperature gradients in the region containing the CJC and terminals.

Thermocouple Linearization. Thermocouple voltage is proportional to, but not linearly proportional to the temperature at the thermocouple connection. There are several techniques for thermocouple linearization. Analog techniques can provide a voltage proportional to temperature from the thermocouple input. In addition, a voltage measurement can be made with an ADC and the temperature looked up in a table, as in the above example. To speed this process, the lookup table can be stored in computer memory and the search performed with an algorithm. Figure 4.03 on the next page illustrates a portion of the NIST lookup table. Finally, thermocouple linearization can be accomplished using a polynomial approximation to the temperature versus voltage curve.

NIST Polynomial Coefficients						
	Type E	Type J	Type K	Type R	Type S	Type T
	Nickel-10% Chromium(+) vs. Constantan(–)	Iron(+) vs. Constantan(–)	Nickel-10% Chromium(+) vs. Nickel–5%(–) (Aluminum Silicon)	Platinum-13% Rhodium(+) vs. Platinum(–)	Platinum-10% Rhodium(+) vs. Platinum(–)	Copper(+) vs. Constantan(–)
	–100°C to 1000°C ±0.5°C 9th order	0°C to 760°C ±1°C 5th order	0°C to 1370°C ±0.7°C 8th order	0°C to 1000°C ±0.5°C 8th order	0°C to 1750°C ±1°C 9th order	–160°C to 400°C ±0.5°C 7th order
a_0	0.104967248	–0.048868252	0.226584602	0.263632917	0.927763167	0.100860910
a_1	17189.45282	19873.14503	24152.10900	179075.491	169526.5150	25727.94369
a_2	–282639.0850	–218614.5353	67233.4248	–48840341.37	–31568363.94	–767345.8295
a_3	12695339.5	11569199.78	2210340.682	1.90002E + 10	8990730663	78025595.81
a_4	–448703084.6	–264917531.4	–860963914.9	–4.82704E + 12	–1.63565E + 12	–9247486589
a_5	1.10866E + 10	2018441314	4.83506E + 10	7.62091E + 14	1.88027E + 14	6.97688E + 11
a_6	–1.76807E + 11		–1.18452E + 12	–7.20026E + 16	–1.37241E + 16	–2.66192E + 13
a_7	1.71842E + 12		1.38690E + 13	3.71496E + 18	6.17501E + 17	3.94078E + 14
a_8	–9.19278E + 12		–6.33708E + 13	–8.03104E + 19	–1.56105E + 19	
a_9	2.06132E + 13				1.69535E + 20	

Temperature conversion equation: $T = a_0 + a_1x + a_2x^2 + \ldots + a_nx^2$

Nested polynomial form: $T = a_0 + x(a_1 + x(a_2 + x(a_3 + x(a_4 + a_5x))))$ (5th order)

Fig. 4.03: Portion of an NIST lookup table

For example, IOtech's DBK19 thermocouple card includes a software driver that contains a temperature conversion library that converts raw binary thermocouple channels, and CJC information into temperature. Included DaqView software provides automatic linearization and CJC for thermocouples attached to the system.

Additional Concerns

Care must be taken when using thermocouples to measure temperature. Sources of minor error can add up to highly inaccurate readings. Additional concerns include:

Thermocouple Assembly. Thermocouples are assembled via twisting wires together, soldering, or welding; all of these techniques, done improperly can introduce errors in the temperature measurement.

Twisting Wires Together. Thermocouple junctions should not be formed by twisting the wires together. This will produce a very poor thermocouple junction with large errors.

Soldering. For low temperature work the thermocouple wires can be joined by soldering, however, soldered junctions limit the maximum temperature that can be measured (usually less than 200 degrees Celsius). Soldering thermocouple wires introduces a third metal. This should not introduce any appreciable error as long as both sides of the junction are the same temperature.

Welding. For higher temperatures, welding is the preferred method of connecting junctions. When welding thermocouple wires together, care must be taken to prevent any of the characteristics of the wire from changing as a result of the welding process. These concerns are complicated by the different composition of the wires being joined. Commercially manufactured thermocouples are typically welded using a capacitance-discharge technique that ensures uniformity.

Decalibration of Thermocouple Wire. Decalibration is another serious fault condition. Decalibration is particularly troublesome because it can result in an erroneous temperature reading that appears to be correct. Decalibration occurs when the physical makeup of the thermocouple wire is altered in such a way that the wire no longer meets NIST specifications. This can occur for a variety of reasons including, temperature extremes, cold-working of the metal, stress placed on the cable during installation, vibration, or temperature gradients.

Insulation Resistance Failure. Temperature extremes can also introduce error because the insulation resistance of the thermocouple will often decrease exponentially as the temperature increases. This can lead to two types of errors: leakage resistance with an open thermocouple and leakage resistance with small thermocouple wire. Figure 4.04 illustrates insulation resistance failure.

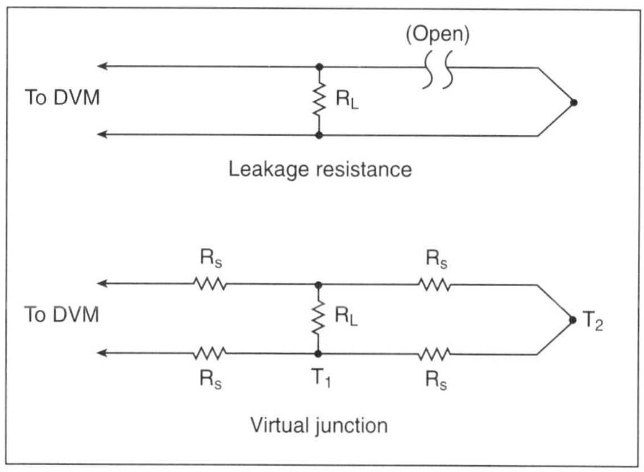

Fig. 4.04: Illustration of a virtual junction resulting from insulation resistance failure

Leakage Resistance with an Open Thermocouple. In high temperature applications, the insulation resistance can degrade to the point where the leakage resistance R_L will complete the circuit and give an erroneous reading.

Leakage Resistance with Small Thermocouple Wire. In high temperature applications using small thermocouple wire, the insulation R_L can degrade to the point where a virtual junction T_1 is created and the circuit output voltage will be proportional to T_1 instead of T_2.

Furthermore, high temperatures can cause impurities and chemicals within the thermocouple wire insulation to diffuse into the thermocouple metal, changing the characteristics of the thermocouple wire. This causes the temperature-voltage dependence to deviate from published values. When thermocouples are used at high temperatures, the atmospheric effects can be minimized by choosing the proper protective insulation.

Other Thermocouple Measurement Issues

For high-accuracy and instrumentation-quality thermocouple measurements, issues such as thermocouple isolation, auto-zero correction, line-cycle noise rejection, and open thermocouple detection must be addressed.

Thermocouple Isolation. Thermocouple isolation reduces noise and the error effects of ground loops. Especially in cases where many thermocouples are bolted directly to a distant metallic object, like an engine block. The engine block and the thermocouple-measurement instrument may have different ground references. Without some form of isolation, a ground loop would be created that would cause the readings to be in error.

Auto-Zero Correction. To minimize the effects of time and temperature drift on the system's analog circuitry, a shorted channel should be constantly measured and subtracted from the reading. Although very small, this circuit drift can become a substantial percentage of the low-level voltage supplied by a thermocouple.

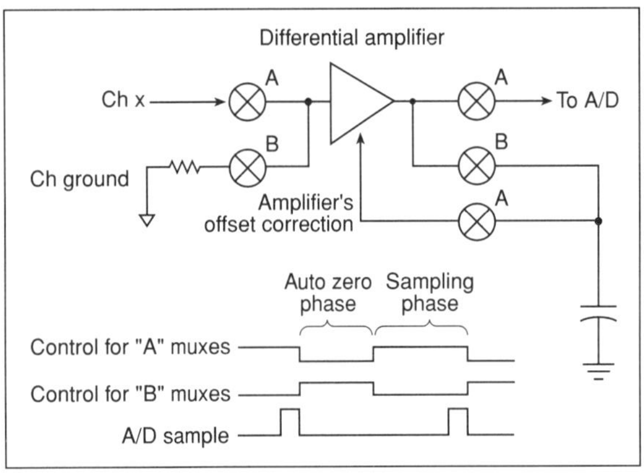

Fig. 4.05: Diagram of auto-zero correction

One method of subtracting the offset due to drift is employed by IOtech's TempScan/1100 high-speed temperature measurement system. Rather than mathematically subtracting the offset resulting from circuit drift, the TempScan/1100 employs a highly accurate analog scheme to automatically subtract the system offset. Before reading any channel, the internal channel sequencer first switches to a reference node and stores the offset error on a capacitor. As the thermocouple channel is switched onto the analog path, the circuit holding the offset voltage is applied to the offset correction input of a differential amplifier, nulling out the offset automatically. Figure 4.05 illustrates this technique.

Line-Cycle Noise Rejection. Because of the relatively small voltage generated by most thermocouples, noise is always an issue. The most pervasive source of noise is from power lines (50 Hz or 60 Hz). Since thermocouples have a bandwidth lower than 50 Hz, a simple per-channel filtering scheme can reduce the incoming AC line noise. For systems with a multiplexed front end, an RC filter is not recommended. Although more complicated, an active filter made up of an op-amp and a few passive components can also be used; however, this method incurs significant additional expense. Furthermore, setup is complicated because each channel needs to be calibrated due to gain and offset error. For these reasons active filters are generally not used. Figure 4.06 is a diagram of an active filter.

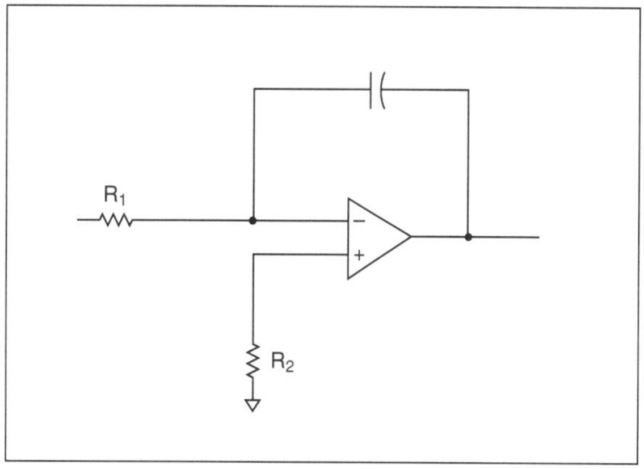

Fig. 4.06: Diagram of active filter

One unique line-cycle averaging technique for reducing noise is employed by the TempScan/1100. A phase-lock-loop circuit monitors and synchronizes the system's internal sampling rate to the multiple of the power line frequency. During each line cycle, 4 channels plus the CJC are sampled 32 times and averaged. This has the effect of reducing the line cycle noise by 82%.

Open Thermocouple Detection. Open thermocouple detection is especially important in systems with high channel counts. Due to age, vibration, and handling, thermocouples can break or become highly resistive. Open thermocouple detection can be accomplished by providing a low level current path to a capacitor across the thermocouple terminals. If the thermocouple is open, the capacitor charges up and drives the amplifier to its rail, indicating a fault. When the thermocouple is properly connected, the low-level current goes through the thermocouple rather than charging the capacitor. Figure 4.07 illustrates an open thermocouple circuit.

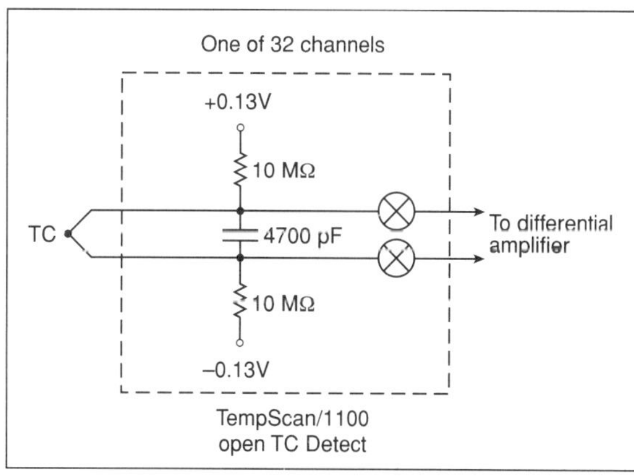

Fig. 4.07: Open thermocouple circuit

Galvanic Action. Some thermocouple insulation contains dyes that form an electrolyte in the presence of water. This creates a galvanic action, with a resultant output hundreds of times greater than the net open circuit voltage. Precautions should be taken to shield thermocouple wires from harsh atmospheres and liquids.

Thermal Shunting. Because it takes energy to heat mass and thermocouples cannot be constructed without mass, the thermocouple will slightly alter the temperature it was meant to measure. The mass of the thermocouple can be minimized using small wire, but a thermocouple made with small wire is far more susceptible to the problems of contamination, annealing, strain, and shunt impedance. To minimize these effects, thermocouple extension wire intended to cover long distances should be used. Commercial thermocouple extension wire is constructed out of material having net open-circuit voltage coefficients similar to specific thermocouple types. It is constructed so that its series resistance is minimal over long distances; it can be pulled through conduit more easily, and it is generally operable over a much lower temperature range than premium-grade thermocouple wire. In addition to offering a practical size advantage, extension wire is less expensive than standard thermocouple wire, especially with platinum-based thermocouples.

Extension wire is generally specified over a narrower temperature range and is more likely to receive mechanical stress, as such the temperature gradient across the extension wire should be kept to a minimum. These simple precautions should ensure that temperature measurements will not be affected by the use of extension wire.

Improving the Accuracy of Thermocouple Wire Calibration. Thermocouple wire is manufactured so that it conforms to NIST specifications. These specifications can often be enhanced by calibrating the wire on site. That is accomplished by testing the wire at known temperatures. (Note: If the wire is calibrated to improve temperature measurements, then it becomes even more imperative that care be taken to prevent decalibration.)

Resistance Temperature Detectors (RTD)

RTDs function on the basic physical principle that a metal's resistance increases with temperature. Most RTDs are simply wire-wound or thin-film resistors made of a wire with a known resistance versus temperature relationship. Platinum is the most commonly used material for RTDs. RTDs are available in a wide range of accuracy specifications; the most accurate RTDs are used as temperature standards at NIST.

Platinum RTDs have resistance values ranging from 10 Ohms for a birdcage model to ten thousand Ohms for a film model. The most common value for RTDs is 100 Ohms at 0°C. Platinum wire has a standard temperature coefficient of α = 0.00385 Ω/Ω/°C, which means a 100 Ohm wire has a temperature coefficient of +0.385 Ω/°C at 0°C; the value for α is actually the average slope from 0°C to 100°C. Chemically pure platinum wire has an α of + 0.00392 Ω/Ω/°C. The slope and the absolute value are relatively small, when you consider the fact that the measurement wires leading to the sensor may be tens of Ohms. With this in mind, even a small lead impedance can cause an appreciable error in temperature measurement.

The following equation states that the change in resistance is equal to α multiplied by the change in temperature.

$$\Delta R = \alpha \Delta T$$

In the above measurement example, a 10 Ohm lead impedance suggests a 10/0.385 @ 26°C error. In addition to the impedance, the lead wire's temperature coefficient can also contribute a measurable error.

To eliminate the possibility of error as a result of additional resistance from the current-carrying excitation leads, an additional set of voltage-sensing leads should be located as close as possible to the sensor. (Recall that the temperature coefficient is 0.39Ω/°C, so a 0.39Ω error in the resistance measurement yields a 1°C error in the measured temperature.) This configuration is called a four-wire RTD measurement because four wires are required between the RTD and the measurement instrumentation. Placing the sense leads at the current supply instead of the RTD provides a less accurate configuration but only requires connecting two wires to the RTD.

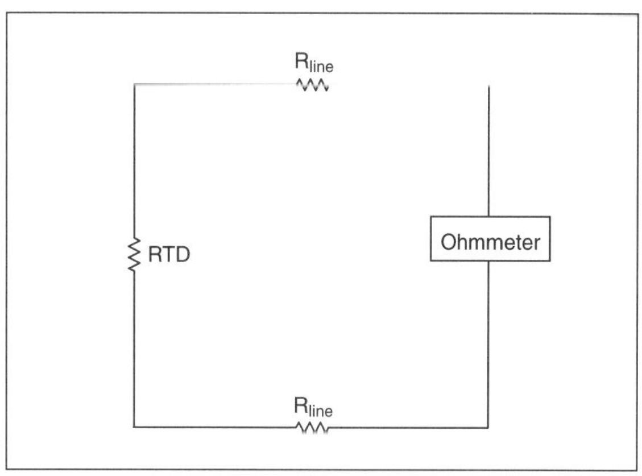

Fig. 4.08: Two-wire RTD measurement

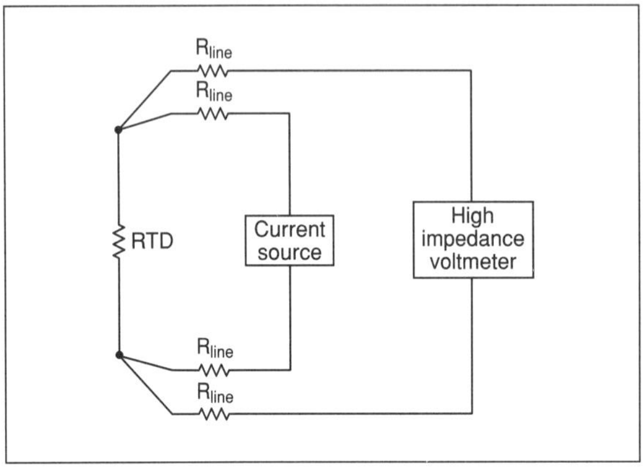

Fig. 4.09: Four-wire RTD measurement with a current source

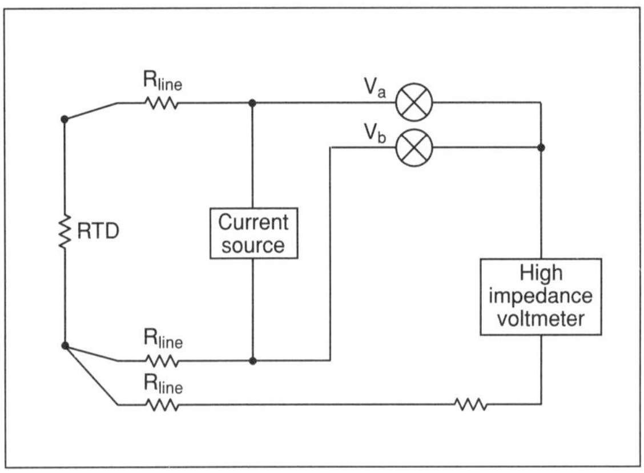

Fig. 4.10: Three-wire RTD measurement with a current source

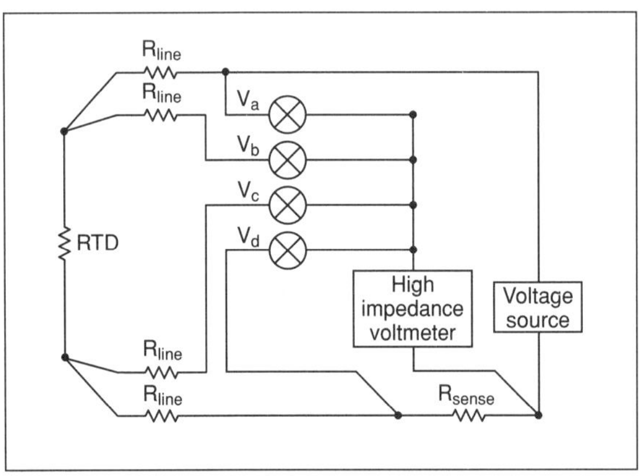

Fig. 4.11: Four-wire RTD measurement using a voltage source

The five common types of circuits, which are used for RTD measurement are shown in Figure 4.08 through 4.12. Following are descriptions of each circuit's operation.

Figure 4.08 (shown on the previous page) shows a simple two-wire resistance measurement. The RTD resistance is read directly from the Ohmmeter. This connection is rarely used since the lead wire resistance must be a known value and it is very difficult to determine the temperature coefficient of the lead wires.

Figure 4.09 shows a four-wire measurement using a current source. The RTD resistance is volts/current. This connection requires a high stability current source and four lead wires.

Figure 4.10 shows a three-wire measurement using a current source. The RTD resistance is derived using Kirchhoff's voltage law as follows: RTD in Ohms = V_a- (2 * V_b)/current. The benefit of this connection over Figure 4.09 is the elimination of one-lead wire. **Note**. This connection assumes that the two-current carrying wires have the same resistance.

Figure 4.11 shows a four-wire measurement using a voltage source. The RTD resistance is derived using Kirchhoff's voltage law as follows: RTD in Ohms = V_a-((V_a - V_b)+(V_c - V_d))/(V_d/R_{SENSE}). The voltage source in this circuit can vary as long as the sense resistor (R_{SENSE}) is stable. This connection requires four-lead wires.

Figure 4.12 shows a three-wire measurement using a voltage source. The RTD resistance is derived using Kirchhoff's voltage law as follows: RTD in Ohms = V_a-((2 * (V_a - V_b))+(V_c - V_d))/(V_d/R_{SENSE}). The voltage source in this circuit can vary as long as the sense resistor (R_{SENSE}) is stable. **Note.** This connection assumes that the two-current carrying wires have the same resistance.

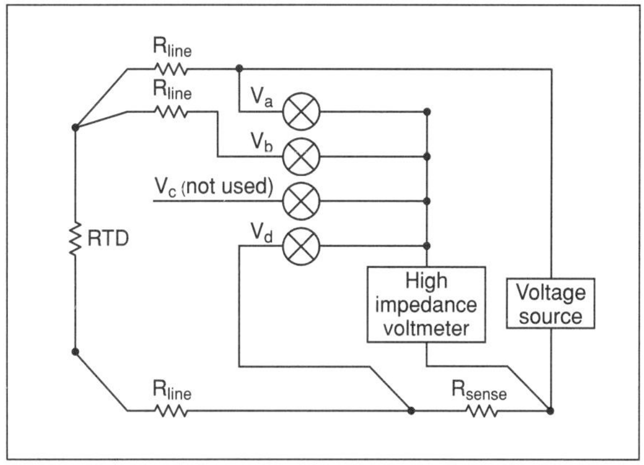

Fig. 4.12: Three-wire measurement using a voltage source

While the RTD is more linear than the thermocouple, its operating range is smaller than a thermocouple's. For maximum accuracy in RTD applications you should use curve-fitting via the Callendar-Van Dusen equation, which follows:

$$R_T = R_0 + R_0\,\alpha\left[T - \delta\left(\frac{T}{100} - 1\right)\left(\frac{T}{100}\right) - \beta\left(\frac{T}{100} - 1\right)\left(\frac{T^3}{100}\right)\right]$$

Where:

R_T = resistance at temperature T
R_0 = resistance at T = 0 C
α = temperature coefficient at T = 0 C
typically +0.00392Ω/Ω/°C
δ = 1.49 (typical value for .00392 platinum)
β = 0, When T>0
0.11 (typical) When T<0

The exact values for coefficients α, δ, and β, are determined by testing the RTD at four temperatures and solving the resulting equations. This Callendar-Van Dusen equation was replaced by a 20th order polynomial to provide a more accurate curve fit. The plot of this equation in Figure 4.13 shows the RTD to be more linear than the thermocouple when used below 800°C (RTDs are not appropriate for applications in which the temperature being measured will exceed 800°C).

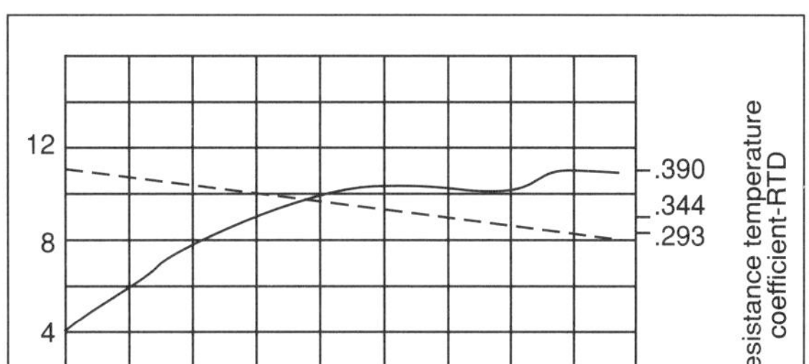

Fig. 4.13: Type S thermocouple vs a Platinum RTD

Another source of error in RTD measurements is resistive heating. A current, I, through a resistance, R, dissipates power $P = I^2R$. For example, a 1 mA current through a 100Ω RTD generates 100 μwatt of power. This may seem insignificant, but it can raise the temperature of some RTDs a significant fraction of a degree. (A typical RTD has self-heating of 1°C/mW. Smaller RTDs offer fast time response, but can have larger self-heating errors.)

Since lower currents generate less heat; currents between 100 and 500 μA are commonly used. This lowers the power dissipation to between 10 and 25 μW, which is tolerable for most applications. Further reducing the current inhibits accurate measurement because as the currents become smaller they are susceptible to noise and more difficult to measure. The RTD's heat can be reduced below 10 μW by switching the current on only when the measurement is made. In a multichannel system, the excitation current can be multiplexed much like the analog inputs. For example, in a 16-channel system, the current will only excite each RTD 1/16th of the time, reducing the power delivered to each RTD from 100% to 6%, (P*1/X where X = the number of channels).

Two practical methods of measuring an RTD include: constant current and ratiometric. An example of a constant current topology can be found in the TempScan/1100's RTD scanning module, the TempRTD/16A (see figure 4.14), which provides a single constant current source of approximately 500 μamps, which is switched among its sixteen channels. A series of front end multiplexers direct the current to each channel sequentially while the measurement is being taken. Both 3- and 4-wire connections are supported to accommodate both types of RTDs. By applying current to only a single RTD at a time, errors due to resistive heating become negligible. Advantages of the constant-current method include circuit simplicity and noise immunity. The disadvantage of using constant current is the expense of building or buying a highly stable current source.

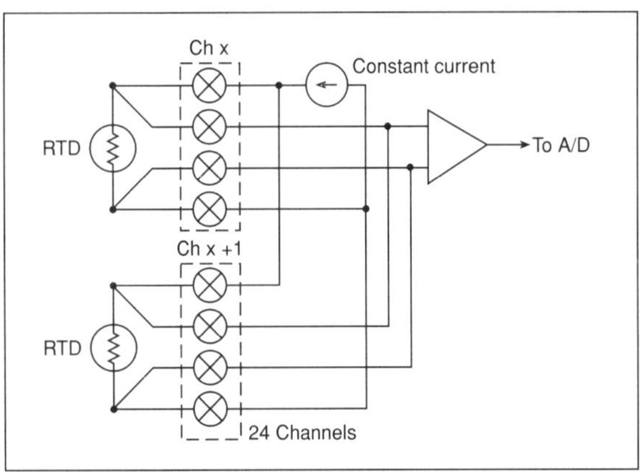

Fig. 4.14: TempRTD/16A card in a TempScan/1100

The DBK9 RTD measurement card for IOtech's DaqBoard, DaqBook, and Daq PCMCIA card PC-based data acquisition products uses the ratiometric method for measuring RTDs. A simple constant voltage source provides a current, I_s, through the RTD and resistor R_d. Four readings, V_a, V_b, V_c, and V_d, are taken for each RTD channel. The current through the RTD, I_s, is V_d/R_d. The voltage across the RTD, V_{rtd}, is equal to V_b-V_c. The RTD resistance is, therefore, V_{rtd}/I_s.

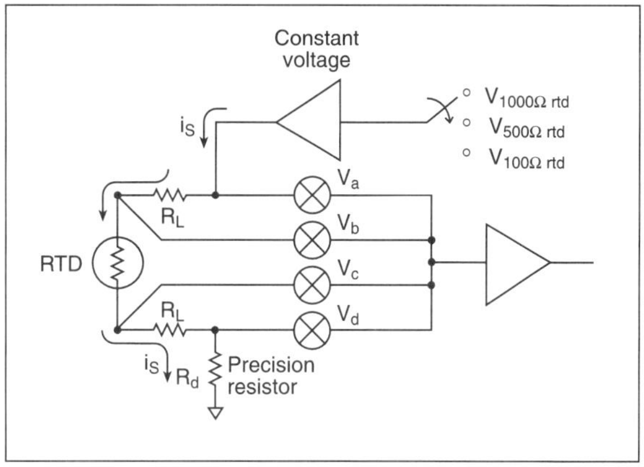

Fig. 4.15: Ratiometric DBK9 front end with 4-wire configuration

For a 3-wire connection scheme, the voltage V_a-V_c includes the voltage drop across only one of the leads. Since the 2 extension wires to the transducer are made of the same metal, one can assume that the drop in the first wire is equal to the drop in the second wire. Therefore, the real voltage across the RTD would be $V_{RTD}=V_a-2(V_a-V_b)-V_d$. The resistance of the RTD can be calculated by $R_{RTD}=R_d[V_{RTD}/V_d]$.

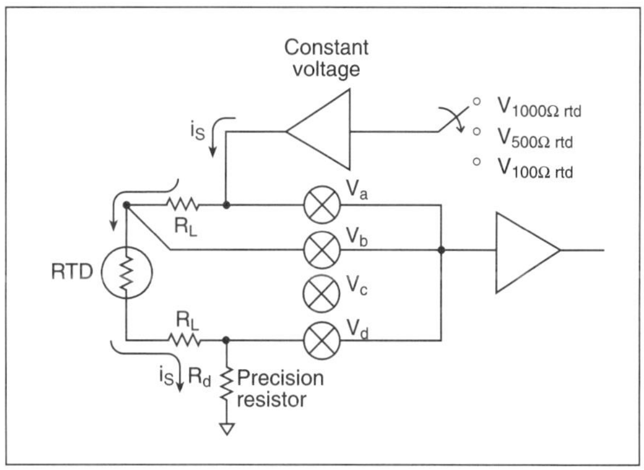

Fig. 4.16: Diagram of DBK9 front end with 3-wire configuration

Practical Precautions

RTDs require the same precautions that apply to thermocouples, including the use of shields and twisted-pair wire, proper sheathing, avoiding stress and steep gradients, and using large extension wire.

In addition, the following issues come into play: the RTD is more fragile than the thermocouple and needs to be protected during use, and RTDs are not self-powered. An excitation current must be provided to the device to obtain a measurement. The current causes Joule heating within the RTD, and this changes the RTD's temperature. This self-heating appears as a measurement error and must be included in the calculations that determine measurement. A typical value for self-heating error is 1°C per milliwatt in free air. An RTD immersed in a thermally conductive medium will distribute this heat to the medium and the resulting error will be smaller. The same RTD that rises 1°C per milliwatt in free air will rise only 1/10°C per milliwatt in air that is flowing at the rate of one meter per second. Using the minimum excitation current that provides the desired resolution and using the largest physically practical RTD will help reduce self-heating errors.

Small Resistance vs. Large Resistance RTDs

	Small RTD	Large RTD
Response Time	Fast	Slow
Thermal Shunting	Low	Poor
Self-heating Error	High	Low

Thermal Shunting. Thermal shunting results in an altered temperature reading as a result of introducing a measurement transducer. Thermal shunting is of greater concern with RTDs because RTDs are larger than thermocouples.

Chapter 5
Strain & Acceleration

This chapter examines the various transducers for measuring strain and acceleration. It also discusses the required signal conditioning, as well as various techniques for ensuring measurement accuracy.

Strain Gages

Strain gages are used in a variety of test and measurement applications. Strain gages can measure most physical phenomena (stress) that elongate or contract a solid material. Strain-gage techniques are also used in many accelerometer and pressure-transducer designs. Strain is defined as the change in length per unit length. For example, if a 1m long bar is stretched to 1.000002 m, the strain is defined as 2 microstrains.

Strain can be measured by bonding a resistive strain gage to the test material. When the test material is strained it transfers that strain to the resistive strain gage, changing its resistance slightly. The "gage factor" is the fractional change in resistance divided by the strain. For example, two microstrain applied to a gage with a gage factor of two causes a fractional resistance change of $4x10^{-6}$. Common gage resistance values are from 120 Ohm to 350 Ohm, but in some applications, resistance can be as low as 30 Ohm or as high as 3000 Ohm.

The Wheatstone Bridge. To make an accurate strain measurement, very small resistance changes must be measured. As in the previous example, if two microstrain are applied to a gage, a change in resistance of only 4 µOhm would need to be measured. The Wheatstone bridge converts a change in resistance into a voltage ready for ADC measurement. If all four resistors are equal, then $V_{out} = 0$. When one of the resistors changes slightly (by a fractional amount X), then a voltage measurable by a strain-gage device is produced.

Full-Bridge Configuration. The optimal strain-gage configuration is the full bridge. This configuration provides the highest sensitivity and the fewest error components. Since the full bridge configuration also provides the greatest output, noise is less of a factor in the measurement. For these reasons, wherever possible, the full bridge configuration is recommended.

A full bridge contains four strain gages mounted on a test member, two on the tensile surface and two on the opposite compressive surface. As the member deflects, two gages increase in resistance while the opposite gages decrease in resistance, maximizing the bridge's output.

In this configuration, the following equation applies: $V_{out} = V_{exc} * X$ where X is the change in resistance (ΔR). Figure 5.01 illustrates a full-bridge configuration.

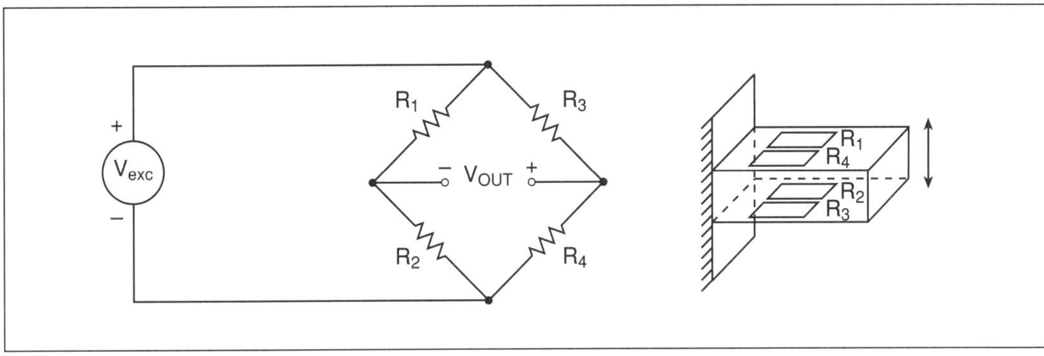

Fig. 5.01: Full bridge and gage locations

Potential error components such as temperature change are nulled by the bridge since all four strain gages have the same temperature characteristics. In a full-bridge configuration, the resistance in the lead wires has no effect on the accuracy of the measurement as long as the input amplifier has a high input impedance. For example, with an input resistance of 100 MOhm, very little current flows through the measurement leads, minimizing the possible voltage drop due to lead resistance.

Half-Bridge Configuration. When physical considerations do not allow for a full bridge, a half bridge can be employed. In this configuration, two strain gages are mounted on a test member while two resistors are used to complete the bridge.

In this configuration, the following equation applies: $V_{bridge} = V_{excitation} * X/2$ where X is approximately the change in resistance (ΔR). For large ΔR, half-bridge and quarter-bridge configurations can introduce an additional nonlinearity error. Figure 5.02 illustrates a half-bridge configuration.

One potential error source with half bridges is that, because of the dissimilar temperature characteristics of bridge completion resistors and strain gages, as their temperature changes, their resistance may not change proportionately. Furthermore, bridge completion resistors are typically not in the same physical location as the strain gages, adding to the error induced by temperature changes.

In systems with long lead wires, it is recommended that the bridge completion resistors be attached close to the gages and not at the measurement device.

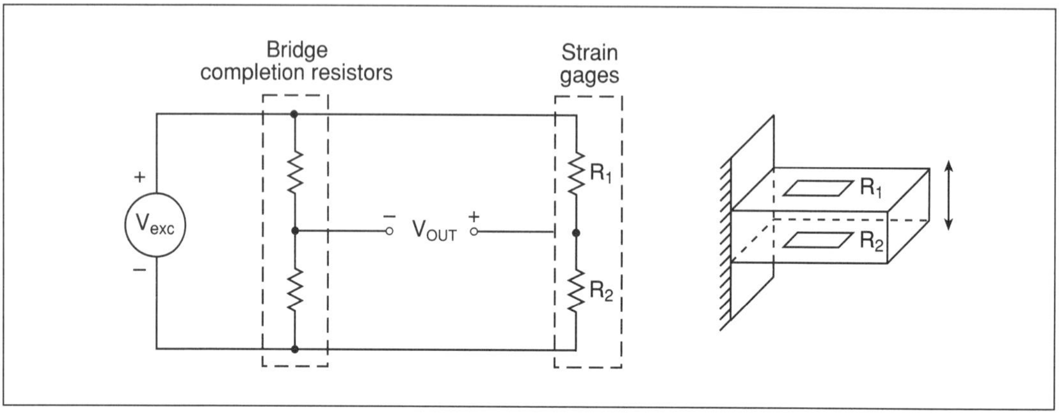

Fig. 5.02: Half-bridge and gage locations

Quarter-Bridge Configuration. A quarter bridge uses one strain gage and three bridge completion resistors. In this configuration, the following equation applies: $V_{out} = V_{exc} * X/4$, where X is approximately the change is resistance (ΔR). Figure 5.03 illustrates a quarter-bridge configuration.

This configuration has the smallest output, making noise a potential problem. Furthermore, all of the error sources and drawbacks in the half-bridge configuration are also applicable to the quarter-bridge configuration.

Excitation Source. Accurate measurement requires a quiet (low noise) and stable regulated excitation source. A regulated excitation source is necessary because the output of a strain gage is proportional to the excitation voltage. Therefore, any fluctuation in the excitation will result in output errors.

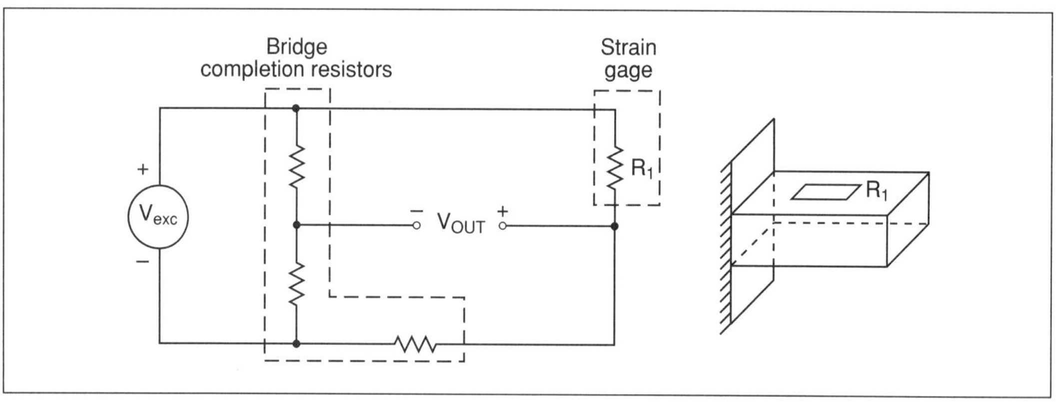

Fig. 5.03: Quarter-bridge and gage location

An example of a strain-gage measurement system is IOtech's DBK43A, an eight-channel signal conditioning option for their 100-kHz PC-based data acquisition systems.

The DBK43A provides one adjustable excitation output per channel. Each channel can be independently adjusted from 1.5V to 10.5V with a current limit of 50 mA. An excitation voltage V, used with a strain gage of resistance R, requires a current of I=V/R.

The composite resistance of a Wheatstone bridge is the same as its components. For example, a 350 Ohm bridge is made up of four 350 Ohm arms. The load current is determined by dividing the excitation voltage by the bridge resistance, in this case 10V/350Ω=0.029A (29 mA).

The amount of Ohmic heating in the strain gages should also be considered because a strain gage's performance can be temperature dependent. In most standard configurations, the dissipated power ($P=V^2/R$) is less than 100 mW in each gage. If the strain gage is bonded to a good heat conductor such as metal, dissipated power is typically not a problem. However, when the medium conducts heat poorly, such as wood, plastic, or glass, care should be taken in choosing the excitation voltage. Care should also be taken when the strain gage is unusually small or when a number of strain gages are placed in proximity.

Finally, one should consider Kelvin excitation. Since the excitation leads carry a small amount of current, there is a voltage drop of $V=IR_{leads}$ between the excitation supply and the strain gages. IOtech's DBK43A, uses Kelvin excitation to measure and stabilize the voltage across the bridge. It accomplishes this by measuring it with a different set of leads than those carrying the current.

If I_e = 50 mA, R_L = 5 Ohms, and the voltage drop in one lead is 250 mV, both exc wires will equal 500 mV. There will be no voltage drop in the sense wire because of negligible current.

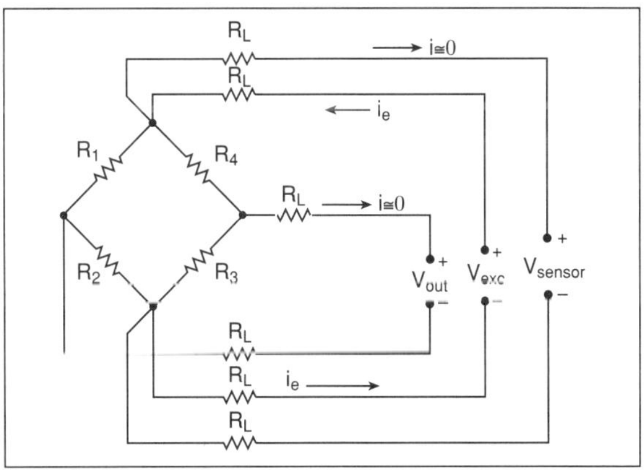

Fig. 5.04: Kelvin strain-gage bridge connection

Strain-Gage Signal Conditioning

Most strain gages and load cells are specified for a full scale value of weight, force, tension, pressure, torque, or deflection with an output of mV/V of excitation. For example, a load cell with a full scale rating of 1,000 lbs might output 2 mV per volt of excitation at full load. For an excitation of 10V, the full scale output is 20 mV.

Steps must be taken to control noise when measuring mV levels. Shielded cables, low-pass filters, differential voltage measurement and averaging can be used to minimize noise in strain-gage measurements.

Reading a mV signal with a 12-bit A/D converter requires an instrumentation amplifier. Most A/D amplifiers have a 10V full scale and thus, only provide a 2.44 mV resolution (10V/4095). To measure strain using the A/D's full resolution, one must adjust the instrumentation amplifier's gain so that the full scale output of the strain gage or load cell covers the entire range of the A/D.

Many test systems are characterized by the presence of a quiescent load that results in an output voltage, even when no force is applied. An adjustable offset can reduce this quiescent output voltage to zero, allowing the applied load to use the full range of the ADC.

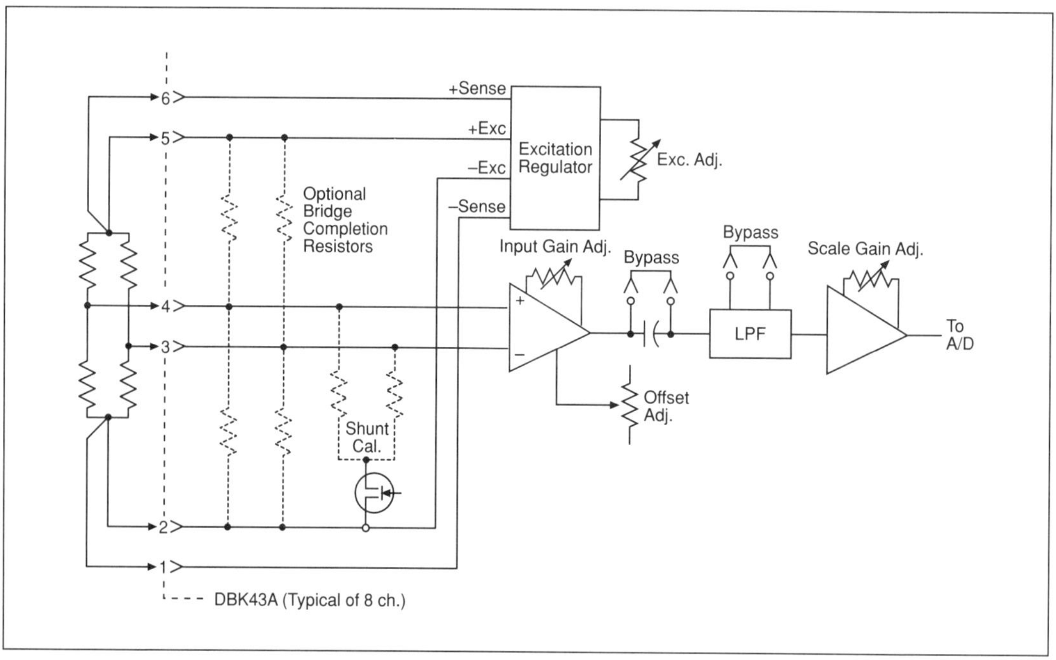

Fig. 5.05: DBK43A strain gage module block diagram

Flexibility is the key to maximizing the performance of a strain gage system. For example, the DBK43A provides adjustable excitation, gain and offset per channel, allowing use of the entire dynamic range of the data acquisition system. Figure 5.05 illustrates one-channel of the DBK43A.

Common Mode Rejection Ratio. Common mode rejection ratio (CMRR) is a vital instrumentation amplifier specification. A strain-gage signal in a Wheatstone bridge configuration is superimposed on a common mode voltage of half the excitation voltage. CMRR measures how well the amplifier rejects common mode voltage. For example, consider a 10V excitation of a strain gage with 2 mV/V output at full scale. An amplifier with a CMRR of 90 dB can introduce an error of 0.158 mV, corresponding to 0.8% full scale. This may not be acceptable. A CMRR of 115 dB is much better, as it introduces an error of only 9 µV, which corresponds to 0.04% of full scale.

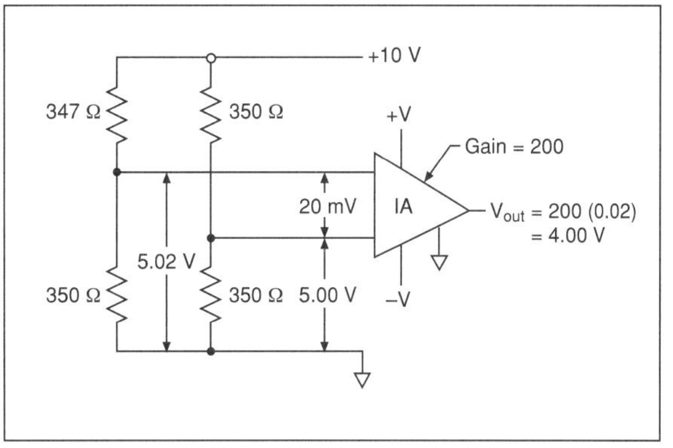

Fig. 5.06: A strain-gage signal in a Wheatstone bridge configuration superimposed on a common mode voltage of half the excitation voltage

$$dB=20\log_{10}(V_{sig}/V_{error})$$
$$V_{max}/V_{error}=\log_{10}{}^{-1}(dB/20)=\log_{10}{}^{-1}(90/20)=31{,}622$$
$$V_{error}=V_{max}/\log_{10}{}^{-1}(dB/20)=5.00/31{,}622=0.158\text{ mV}$$
$$\%error=(V_{erro}/V_{sig})100=0.158\text{ mV}/20\text{ mV}(100)=0.79\%$$

IOtech's DBK43A provides a regulated excitation source with optional Kelvin excitation. Provision is made for on-board bridge-completion resistors when using quarter- and half-bridge strain gages. Two instrumentation amplifiers provide input and scaling gain adjustments. An offset adjustment permits nulling of large quiescent loads, so that the input signal uses the full range of the data acquisition system. This allows the measurement to have the full resolution of the ADC.

Many strain-gage signal conditioners provide fixed gain, offset, and excitation settings; typically the manufacturer suggests that the user adjust the system via software. The problem with this solution is that it's generalized; fixed settings do not take advantage of the maximum dynamic range of the A/D. This decreases the actual available resolution of the measurement.

For example, many generic strain-gage signal-conditioning modules can be configured with a fixed 3mV/V rating. At 10V, the excitation, offset, and gain trimming are fixed and no adjustments are available.

The excitation adjustment provides a means for the user to set the excitation to the maximum allowed by the manufacturer, maximizing the bridge's output. The offset adjustment allows the user to zero the output offset caused by a slight bridge imbalance or quiescent deformation of the mechanical member. The gain adjustments allow the user to set a gain that will provide a full scale output under maximum load, optimizing the dynamic range of the A/D.

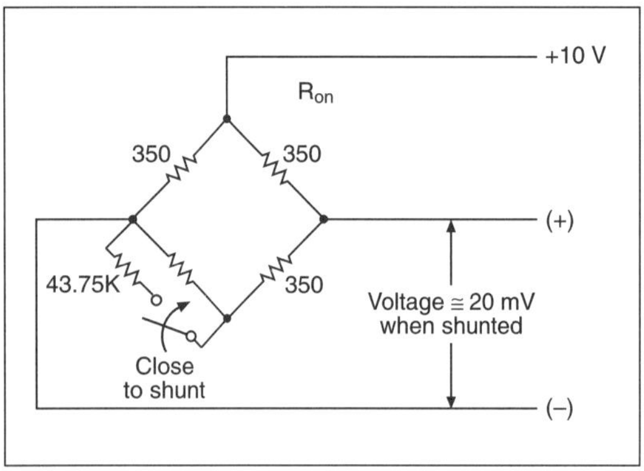

Fig. 5.07: Schematic of a shunt bridge

The DBK43A provides a shunt calibration feature, illustrated in Figure 5.07, that allows the user to switch user-supplied shunt resistors into either of the two lower legs of the bridge via software. This is useful because in some instances, it is not possible to provide a maximum load to set the optimum gain. A shunt resistor value can be calculated so that when it is switched in, it will simulate a full load, allowing an accurate gain to be set. For any balanced bridge, there is a resistance value that can be applied in parallel with one of the four bridge elements to create predictable imbalance and output voltage. For example, a 350 Ohm 2 mV/V strain gage will deliver full output if one leg drops by 0.8% (about 2.80 Ohm) to 347.2 Ohm. A 44K Ohm resistance shunted across one or the other lower bridge elements will result in full positive or full negative output.

An appropriate formula for the shunt-cal resistance value is:

$$R_{shunt} = R_{bridge} \text{ arm } [V_{excitation} / 4 (V_{out})]$$

For example:

$$R_{shunt} = 350 [10 / 4 (0.020)] = 43{,}750 \text{ Ohm}$$

Many products provide calibration software. The DBK43A ships with GageCal, a Windows-based program that provides three calibration methods, on-line instructions, and a diagnostic screen for testing the calibrated system.

Load Cells and Torque Sensors

Strain gages are commercially available in prefabricated modules such as load cells, configured to measure force, tension, compression, and torque. Load cell manufacturers provide calibration and accuracy information. Load cells typically use a full-bridge configuration and have four leads for bridge excitation and measurement.

Strain Diaphragm Pressure Gages. A strain diaphragm pressure gage consists of two or four strain gages on a thin diaphragm. The strain gages are wired in a Wheatstone bridge configuration, including bridge completion resistors if necessary, so that the pressure gage is electrically equivalent to a load cell. The output voltage is specified as a millivolts per volt of excitation for a full scale pressure differential across the diaphragm.

If the reference pressure is open to air, the gage compares the inlet pressure to the ambient pressure. Ambient pressure is approximately 14.7 lbs/in^2 at sea level. If the gage is going to be used to measure ambient pressure, it is necessary to use a sealed reference chamber with either vacuum reference pressure (0 lbs/in^2) or sea-level reference pressure (14.7 lbs/in^2).

Temperature variations can affect the accuracy of strain diaphragm pressure gages. A pressure gage with a sealed nonzero reference pressure exhibits temperature variations in a manner consistent with the ideal gas law. Near room temperature, for example, a 5°C change in the ambient temperature causes an error of 1.7% in the pressure measurement. Temperature variations can also affect the performance of the strain gages themselves. Accurate pressure measurements in environments with varying temperatures require transducers that compensate for such effects.

All strain diaphragm pressure gages require a regulated excitation source. Some gages provide regulation internally, requiring the user to obtain an unregulated voltage from a power supply. Some strain diaphragm pressure gages also employ internal signal conditioning, which amplifies the millivolt signal from the Wheatstone bridge to a full scale voltage of 5V to 10V; gages of this type have low-impedance output. In contrast, other pressure gages have no internal signal conditioning, so their output impedance is equal to that of the Wheatstone bridge (several KOhm for semiconductor strain gages), and their full scale voltage is millivolts. In short, one must carefully consider the kind of pressure gage to use and be prepared to provide the necessary excitation and signal conditioning.

Common Parameters to Describe a Physical System. Two parameters used to describe a physical system are the "Q" factor and the natural resonant frequency W_0. The Q factor is the measure of a system's capacity to either absorb or deliver energy per cycle of excitation. For applications that are lightly damped, such as a piezoelectric accelerometer (discussed in detail later), the maximum response is very close to W_0 and the amplification factor is Q.

When comparing the frequency response of different applications, the frequency axis is often normalized with respect to the system's resonant frequency, W_o. This normalization facilitates qualitative comparisons and mathematical operations.

Strain-Gage Accelerometers. Strain-gage accelerometers use a test mass attached to a cantilever. The strain gage measures the bend of the cantilever, which is a result of acceleration. These accelerometers typically have natural frequencies of several kilohertz. If a system has no additional damping, the cantilever-mass combination forms a harmonic oscillator with a very large Q, near 10^5.

A system with a high Q, excited near its resonant frequency will produce large oscillations. This leads to inaccurate measurements, and may even result in the accelerometer being ripped apart. Consequently, oil is sometimes added to increase the damping. Figure 5.08 and figure 5.09 show the frequency response of harmonic oscillators with Q = 1 and Q = 10^5. For Q = 10^5, the accelerometer is usable with 10% accuracy up to 1/3 of its resonant frequency. In contrast, for Q = 1, the accelerometer is usable with 10% accuracy to 1/2 of its resonant frequency.

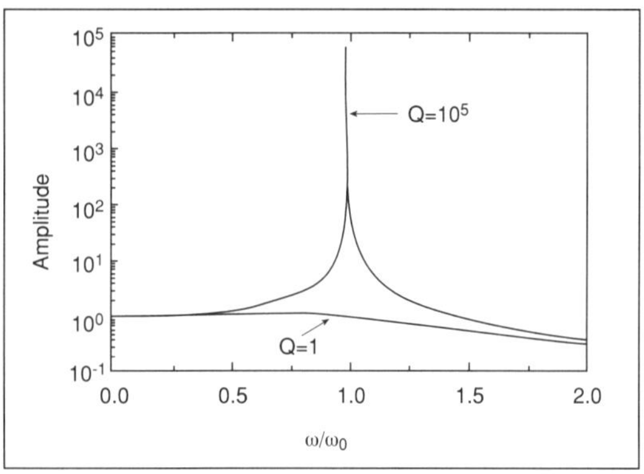

Fig. 5.08: Frequency response of typical harmonic oscillator

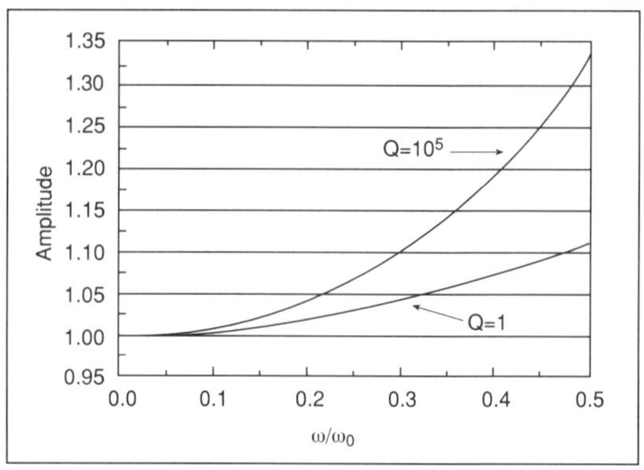

Fig. 5.09: Frequency response illustrated in Fig. 5.08, magnified to show the low frequency response

Most strain-gage accelerometers have strain gages wired in a Wheatstone-bridge configuration. Signal conditioning for a common strain-gage bridge would apply to this type of transducer since it looks exactly like a load cell.

Piezoelectric Transducers

When subjected to a dynamic mechanical force, piezoelectric crystals generate an electric charge on opposite faces. A variety of transducer types use piezoelectric crystals to convert mechanical quantities into electrical signals. Quartz is the most commonly used material in Piezoelectric transducers (PZT). Quartz crystals that respond to either compression or shear forces are readily available.

From an electrical point of view, a PZT looks like a capacitor with a time varying charge, Q(t). This charge is proportional to the force on the crystal and is usually measured using either voltage or charge amplification.

Voltage Amplifier. In Figure 5.10, the voltage amplification gain is 1. Adding feedback resistors or increasing the number of amplification stages can result in different gains. The amplifier converts the high-impedance voltage input to a low-impedance voltage output. The voltage is Q*C, where Q is the charge generated across the PZT, and C is the total capacitance over which this charge is distributed (including the PZT and cable capacitance). Both the cable and PZT capacitance may be on the order of several hundred picofarads (pF). As a result, changing the cable can change the PZT's sensitivity. The best implementation of this configuration places the op amp close to the PZT. Doing so minimizes cable capacitance and maximizes sensitivity. (**Note:** The length of the cable beyond the amplifier does not affect sensitivity.)

Charge Amplification. The charge amplifier configuration shown in Figure 5.10 is extremely versatile because its output voltage is $V = Q/C$. Q represents the charge generated by the PZT, and C is the feedback capacitance. If the op amp has a very large open-loop gain, as most op amps do, the output voltage will remain independent of the cable capacitance. This allows the amplifier to be placed a moderate distance (several meters) from the PZT. However, since noise increases with cable length, cable length limits are set by noise requirements as opposed to sensitivity considerations.

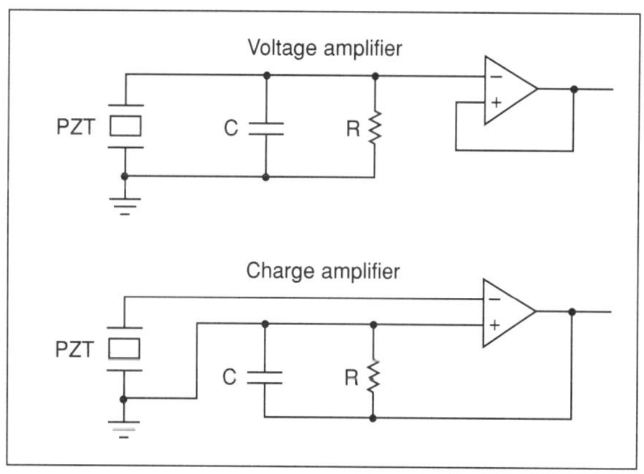

Fig. 5.10: PZT voltage and charge amp

The charge amplifier acts like a high-pass filter with lower corner frequency $f = 1/(2\ RC)$, creating a trade-off between sensitivity and frequency response. Reducing C increases sensitivity, but it also increases the lower corner frequency. (**Note:** In the charge amplifier configuration, PZTs can only be used for dynamic measurements.) The leakage resistance of most PZTs prevents static measurements, as most PZTs fall in the 10^{+10} to 10^{+12} Ohm range. Such leakage resistance, combined with a capacitance of several hundred pF, gives a time constant on the order of a few seconds.

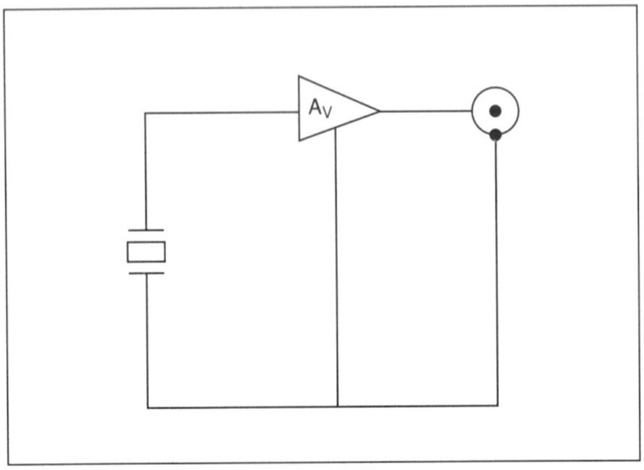

Fig. 5.11: Block diagram of a typical ICP®

Low-Impedance Transducers. Most modern PZTs feature built-in signal-conditioning circuitry. Such units have a low-impedance output and require an external power supply. Also known as integrated circuit piezoelectric (ICP®) transducers, their sensitivities are specified by the manufacturer, so the user's only concerns are interfacing the PZT to standard voltage-measuring equipment, and converting the data from voltage to engineering units.

The most common application of low-impedance PZT transducers is in accelerometers. Accelerometers are used to measure acceleration and mechanical vibration in a variety of applications such as automobile safety systems, which use accelerometers to deploy airbags; the space shuttle uses accelerometers to monitor lift-off acceleration and motion during orbit, and many industrial processes use accelerometers to ensure vibration in machinery remains below acceptable levels.

All accelerometers use a mass to convert acceleration to force via the relationship $F = ma$. This force is converted to an electric signal using a strain gage or PZT. The strain-gage technique permits static acceleration measurements, whereas PZT accelerometers do not. These two common types of accelerometers resemble simple harmonic oscillators.

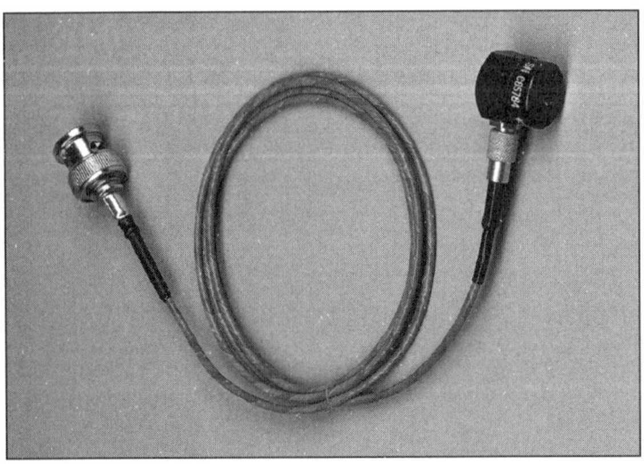

Fig. 5.12: ICP® Accelerometer

ICP is a registered trademark of PCB Piezotronics, Inc.

A low-impedance Piezoelectric accelerometer consists of a Piezoelectric crystal and an electronic amplifier. When stretched or compressed, the crystal develops a charge variation between its two surfaces that is related to the amount of stress, shock, or vibration on the crystal. The amplifier transforms the sensor's high impedance to the output impedance of a few hundred Ohms. Low-impedance Piezoelectric transducers are used to measure pressure and force, as well as acceleration. The accelerometer circuit requires only two wires (coax or twisted pair) to transmit both power and signal. At low impedance, the system is insensitive to externally induced or "triboelectric" cable noise. Sensitivity is not affected by cable length.

Piezoelectric accelerometers are composed of a Piezoelectric crystal attached to a seismic mass. Acceleration causes the seismic mass to exert a force on the PZT, generating an electrical signal. The seismic mass is attached to a bolt that acts like a stiff spring, giving the system a higher resonant frequency than Piezoresistive accelerometers, typically 20 kHz. Piezoelectric accelerometers with resonant frequencies as high as 120 kHz are available, giving them a usable frequency range from less than 1 Hz to 40 kHz.

Like other PZTs, Piezoelectric accelerometers are available in both low-impedance versions, which must be powered externally, and in high-impedance versions, which are self-generating. A vibrating test fixture introduces yet another noise source for high-impedance accelerometers. If the cable between the vibrating accelerometer and the signal conditioning circuit moves, the shield may rub against the insulator in the cable, generating a charge that produces additional noise. This is called the triboelectric effect.

Due to their high noise immunity and simple connectivity, low-impedance ICP accelerometers are most commonly used today. These transducers need only a constant current source for excitation.

Figure 5.13 shows a simplified accelerometer-to-signal-conditioning card connection. In this case the signal conditioning card is IOtech's DBK4 dynamic signal-input card. The voltage developed across R is applied to the gate of the metal-oxide-silicon-field-effect transistor (MOSFET). This MOSFET is powered from a constant current source of 2 or 4 mA.

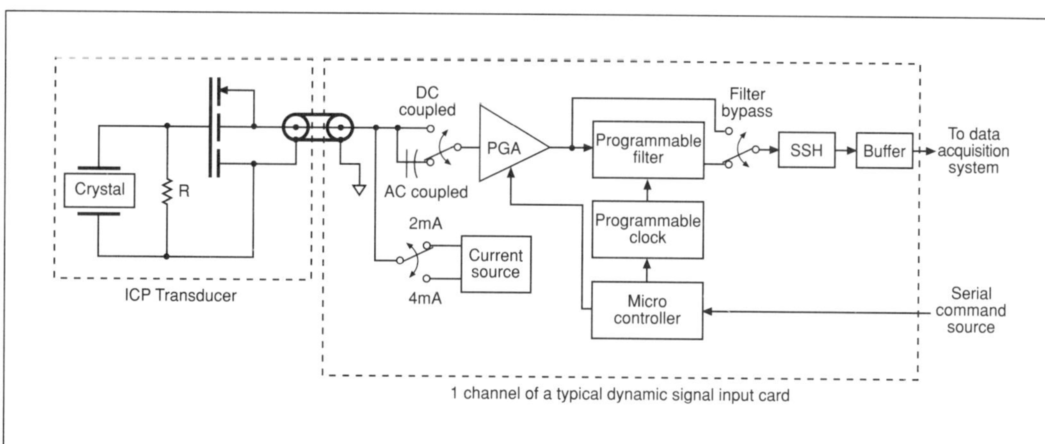

Fig. 5.13: ICP® accelerometer attached to a typical dynamic signal-input card

The MOSFET circuit will bias off at approximately 12V in the quiescent state. As the system is excited, voltage is developed across the crystal and applied to the gate of the MOSFET. This voltage will cause linear variation in the impedance of the MOSFET, which will cause a proportional change in bias voltage. This voltage change will be coupled to the DBK4 input amplifier through the capacitor C. The low-frequency corner is controlled by the value of R and the internal capacitance of the Piezoelectric crystal. Units weighing only a few hundred grams can provide high level outputs up to 1 V/g with response to frequencies from 0.3 Hz to 2 kHz. Smaller units with less sensitivity respond to frequencies from 1 Hz to 35 kHz.

The constant current source provides a source-gate bias for the FET. As the gate current of the FET changes as a result of pressure on the crystal, the voltage drop, V_{ds}, changes proportionately. This voltage is the output of the pre-amp. An AC coupling circuit, or high-pass filter, is always necessary because there is high DC offset present at V_{ds} which is a result of the bias current. The cutoff frequency of the high-pass filter is dependent on the application and the particular accelerometer. Figure 5.14 is a constant current source schematic.

To eliminate the necessity of an out-board pre-amp, some accelerometer input options have the current source and the AC coupling circuitry built-in, providing direct BNC connection between the accelerometer and the data acquisition system.

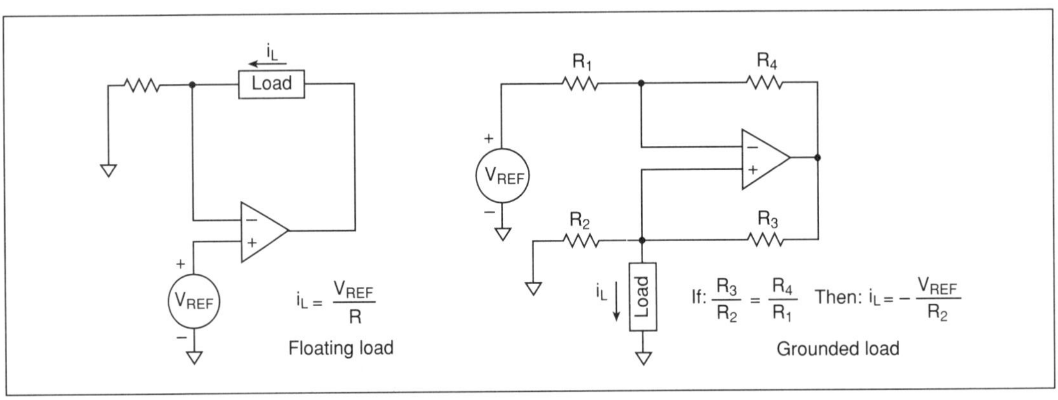

Fig 5.14: Constant current source schematic

Most accelerometers provide an output that needs to be amplified and filtered prior to A/D conversion. A programmable gain amplifier is best suited for accelerometer applications so that the gain can be interactively adjusted by the operator for the optimal response.

Along with programmable amplification, programmable low-pass filters are highly recommended to reject unwanted high-frequency signals. These signals are typically due to noise or higher frequency vibration components that are not germane to the application. When developing the front-end circuitry for this type of measurement, noise rejection and bandwidth are of primary concern. As the bandwidth increases, the noise can increase as well.

Low-pass filtering should be used in most accelerometer measurements to reduce noise and aliasing effects. The cutoff frequency of the low-pass filter should be close to the test's maximum useful operating frequency.

Chapter 6
General Amplification

General Amplification

Data acquisition systems use many amplification circuits to accomplish their tasks. The versatility of operational amplifiers makes them the universal analog building block for signal conditioning. In this chapter, some uses of amplifiers in data acquisition equipment are discussed.

Data Acquisition Front Ends

Data acquisition systems differ from single channel measurement instruments because they are capable of measuring multiple channels. A voltmeter with a selector switch to allow measurement of multiple voltages is a very simple data acquisition system. The high degree of operator interaction required limits the performance of this simple system, but the function is the same.

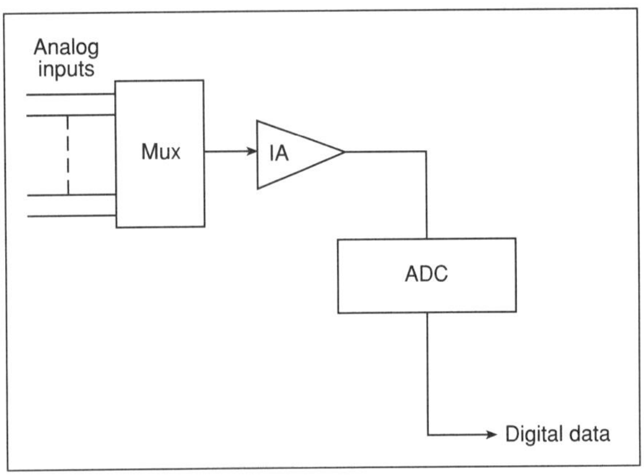

Fig. 6.01: Simple data acquisition system

Figure 6.01 illustrates a simple data acquisition system which consists of a switching network (MUX), an instrumentation amplifier (IA) and an analog-to-digital converter (ADC). These three sections allow measurement of multiple voltage channels using a single A/D. The individual circuit blocks each have unique capabilities and limitations which together define the performance of the system.

Switching Networks

The switching network in an automated data acquisition system can be either relay or solid-state, depending primarily on channel-to-channel switching speed and isolation requirements. In high-speed systems, defined as those switching channels more than 200 times per second, solid state switching is used. Solid state switching networks are configured with analog switches or multiplexers and are capable of switching channels over 1,000,000 times per second. Some drawbacks to solid state switches are input voltage limitations, on-resistance, charge injection, and crosstalk.

Input Voltage Limitations. Input voltage limitations of analog switches are generally defined by the connected power supply voltages, usually plus and minus 15V measured with respect to analog common. There are also "protected" devices which can withstand somewhat higher voltages, up to 30-50 volts. The voltage limitations of analog switches and multiplexers also depends on whether or not the devices are powered at the time the input voltages are applied. As a general rule, it is better to have the analog data acquisition system powered before connecting to live measurement voltages.

On-resistance. All switches have on-resistance. An ideal switch, which is closely approximated by a mechanical switch or relay contact, has zero resistance when the switch is closed. A reed relay contact can be less than 10 milli-ohms. A good analog switch can have on-resistance of 10-100 Ohms, while an analog multiplexer can have 100-2500 Ohms of on-resistance per channel. The on resistance adds directly to the signal source impedance and can affect measurement accuracy.

Charge injection. Charge injection is an undesirable characteristic of analog switching devices, which couples a small portion of the input gate drive voltage to the analog input and output signals. From the standpoint of the measurement, this glitch in the output signal can cause error and will be measured on the input signal if the source impedance is not low. There are circuit design techniques which can be used to minimize the effects of charge injection, but the most effective method is to keep source impedance as low as possible.

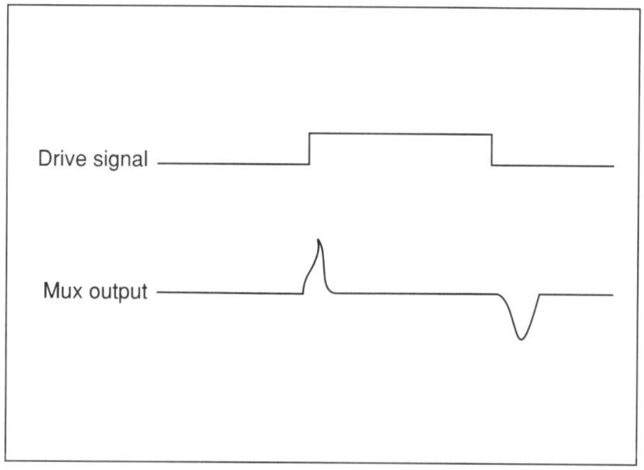

Fig. 6.02: Effect of charge injection on a mux output

Channel-to-channel Crosstalk. Channel-to-channel crosstalk is another non-ideal characteristic of analog switching networks, especially IC multiplexers. Crosstalk occurs if the voltage applied to any one channel affects the accuracy of the reading from another channel. Conditions are especially optimum for crosstalk when signals of relatively high magnitude and high frequency are being measured, such as 4V to 5V signals connected to channel one while 100 mV signals are on the other channels. Multiplexing at high frequencies also exacerbates crosstalk because of capacitive coupling between switch channels. Again, the best defense from crosstalk is keeping source impedances low.

Instrumentation Amplifiers

The block following the switching network in a data acquisition system is an amplification block. This stage uses an amplifier configuration called an instrumentation amplifier and has several important functions. The functions are: rejection of common-mode voltages, signal voltage amplification, minimizing the effect of multiplexer on resistance, and properly driving the ADC input.

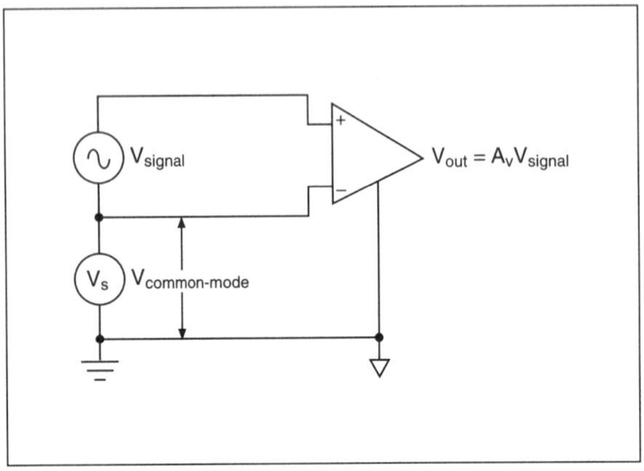

Fig 6.03: Typical instrumentation amplifier

Common mode voltage is the voltage measured relative to analog common which is common to both input voltages. The two input voltages could be 4.1V and 4.2V, for example; the common mode voltage is 4.1V and the differential voltage between the two is 0.1V. Ideally the instrumentation amplifier (IA) ignores the common mode voltage and amplifies the difference between the two inputs. The degree to which the amplifier rejects common mode voltages is given by a parameter called common-mode rejection ratio (CMRR). The CMRR is expressed in decibels (dB's); a ratio of 10 is 20 dB and a ratio of 100 is 40 dB. Each successive ratio of 10 adds 20 dB. If an instrumentation amplifier with a CMRR of 80 dB is measuring a 20 mV signal in the presence of a 4V common mode voltage, the error due to the common mode voltage would be 0.4 mV, which is still 2% of the signal. The ability of an IA to reject "high" common mode voltages is sometimes misinterpreted as an ability to reject high voltages. What is really being rejected is a voltage level having a high magnitude relative to a small signal voltage, such as 4V relative to 20 mV.

Signal voltage ranges are frequently smaller than the range of the system ADC. For example, a 0-100 mV signal is much smaller than the 0-5V range of a typical ADC. To achieve the maximum practical resolution for the measurement, a gain of 50 is needed. Instrumentation amplifiers are capable of gains in this range. In fact, instrumentation amplifiers are capable of gains ranging from unity to well over 10,000 in some specific applications. In multiplexed systems, the gains of IA's are usually restricted to the range of 1-1,000.

There is a special class of instrumentation amplifiers with programmable gain which can switch between fixed gain levels at sufficiently high speeds to allow different gains for different input signals delivered by the input switching system. These amplifiers are called programmable gain instrumentation amplifiers (PGIA). The same digital control circuitry which selects the input channel can also select a gain range allowing a high degree of flexibility in the system.

The non-ideal on-resistance of analog switches will add to the source impedance of any signal source to create measurement error. This effect is minimized by the extremely high input impedance of the IA. The input stage of an IA consists of two voltage followers, which have the highest input impedance of any common amplifier configuration. The high impedance and extremely low bias current drawn from the input signal create a minimal voltage drop across the analog switch sections resulting in a more accurate signal reaching the input to the IA.

The output of the instrumentation amplifier has a low impedance which is ideal for driving the ADC input. The typical ADC does not have high or constant input impedance. It is important for the preceding stage to provide a signal with the lowest impedance practical.

Instrumentation amplifiers have some limitations including offset voltage, gain error, limited bandwidth, and settling time. The offset voltage and gain error can be calibrated out as part of the measurement, but the bandwidth and settling time are parameters which limit the frequencies of amplified signals and limit the frequency at which the input switching system can switch channels between signals. A series of steady DC voltages applied to an instrumentation amplifier in rapid succession creates a difficult composite signal to amplify. The settling time of the amplifier is the time necessary for the output to reach final amplitude to within some small error (often 0.01%) after the signal is applied to the input. In a system scanning inputs at 100 kHz, the total time spent reading each channel is 10 μS. If analog-to-digital conversion requires 8 μS, settling time of the input amplitude to the required accuracy must be less than 2 μS.

Although it has been stated that offset voltage and gain error can be calibrated out, it is important to understand the magnitude of the problem to allow good decisions as to when these error corrections are necessary. An IA with an offset voltage of 0.5 mV and a gain of X2 measuring a 2V signal will only incur an error of 1 mV in 4V on the output which is 0.025%. An offset of 0.5 mV and a gain of X50 measuring a 100 mV signal will incur an error of 25 mV in 5 volts or 0.5%. Gain error is similar. A stage gain error of 0.25% will have a greater overall effect as gain increases resulting in larger errors at higher gains and minimal errors at unity gain. System software can generally handle known calibration constants with mx + b routines but some measurements are not critical enough to justify the effort.

Analog-to-Digital Converters

The analog to digital converter stage is the last link in the chain between the analog domain and the digitized signal path. In any sampled data system, such as a multiplexed data acquisition system, there is a functional need for a sample-hold stage preceding the ADC. The ADC cannot correctly digitize a time varying voltage to the full resolution of the ADC unless the voltage changes very slowly. Some ADC's have internal sample-holds or use architecture which emulates the function of the sample-hold stage. For the purpose of this discussion, it is assumed that the ADC block includes a suitable sample-hold circuit to stabilize the input signal during the conversion period.

The primary parameters of ADC's in data acquisition systems are resolution and speed. Data acquisition ADC's typically run at speeds from 20 kHz to 1 MHz and have resolutions of 12- or 16-bits. ADC's have one of two types of input voltage range, unipolar or bipolar. The unipolar type range typically runs between 0 volts and some positive or negative voltage such as 5V. The bipolar type range will typically run from a negative voltage to a positive voltage of the same magnitude. Many data acquisition systems have the capability of reading bipolar or unipolar voltages to the full resolution of the ADC, which necessitates some sort of level shifting stage to allow bipolar signals to use unipolar ADC inputs and vice versa. A typical 12-bit, 100-kHz ADC has an input range of -5V to +5V and a full scale count of 4,096. Zero volts corresponds to nominally 2,048 counts. If the range of 10 volts is divided by the number 4,096, the LSB (least significant bit) magnitude of 2.44 mV is obtained. A 16-bit converter with the same range and 65,536 counts has an LSB value of 153 µV.

Signal Conditioning Amplification

Many real world sensors have very low signal levels. These levels are too low for direct application to multiplexed data acquisition system inputs with relatively low gains. Two common examples are thermocouples and strain-gage bridges which both deliver full scale outputs of less than 50 mV. There are two broad classes of signal conditioning circuits, multiplexed and dedicated. In multiplexed conditioners, multiple input sources are sequentially switched through one amplifier signal path allowing for an economical cost per channel with a somewhat reduced performance. The more expensive dedicated type have amplifier stages for each signal path and multiplex the high-level signals only.

Most amplifier stages are applications of one of two basic configurations, inverting or non-inverting. See figure 6.04 for an illustration of each. Each configuration has a simple equation which provides the idealized circuit gain as a function of the input and feedback resistors. There are also special cases of each configuration which make up the rest of the fundamental building blocks, namely the unity gain follower and the difference amplifier.

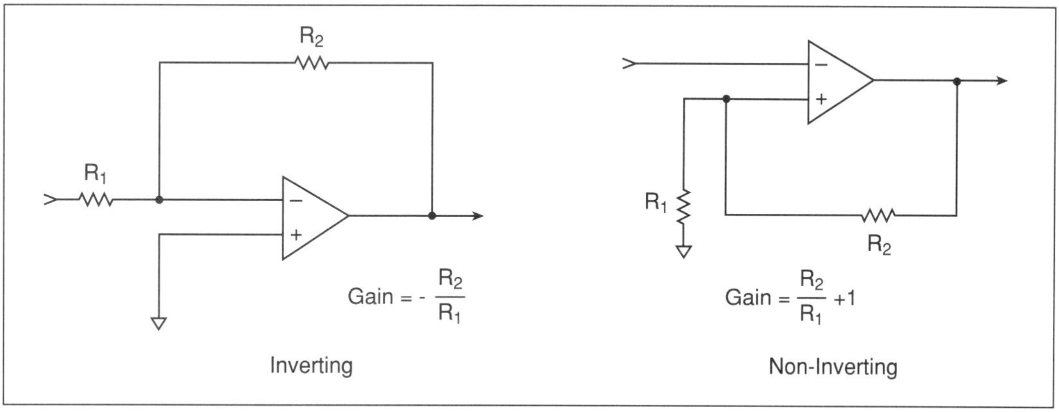

Fig. 6.04: Inverting and non-inverting amp

Source Impedance & Multiplexing

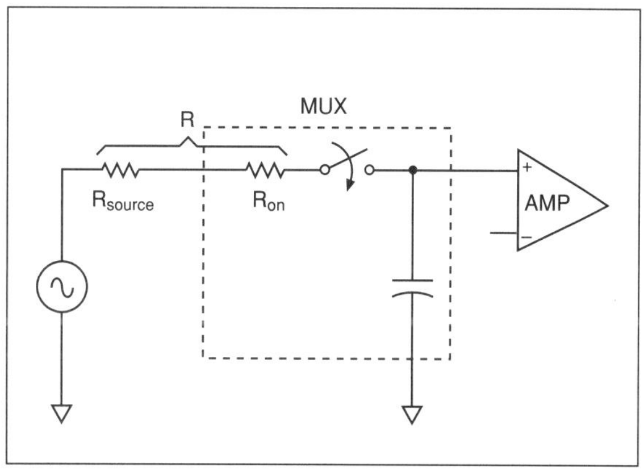

Fig. 6.05: Simple RC circuit

Multiplexing and a high source impedance do not mix well. The reason that low source impedance is always important in a multiplexed system can be explained with a simple RC circuit, see figure 6.05. Multiplexers have a small parasitic capacitance to analog common on all signal inputs as well as outputs. These small capacitance values, when combined with source resistance and fast sampling time, affect measurement accuracy. A simple RC equivalent circuit consists of a DC voltage source with a series resistance, a switch, and a capacitor. When the switch closes at t = 0, the voltage source charges the capacitor through the resistance. When charging 100 pF through 10 kOhm, the RC time constant is 1 µS. In a 10 µS time interval of which 2 µS is available for settling time, the capacitor will only charge to 86% of the value of the signal, introducing a major error. If the same calculations are done with a 1,000 Ohm resistor, the capacitor will easily charge to an accurate value in 20 time constants.

Voltage Measurement Errors

One must consider several factors when making voltage measurements. This section will discuss a few of these, including: input and source impedance, differential voltage measurements, and isolation considerations.

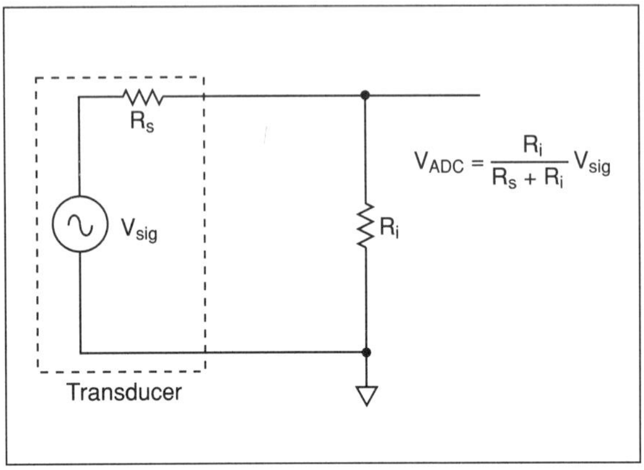

Fig. 6.06: Input and source impedances

Input and Source Impedance

Figure 6.06 depicts how system input impedance and the transducer's source impedance combine to form a voltage divider, which reduces the voltage being read by the ADC. The input impedance of most input channels is 1M Ohm or greater, so it is usually not a problem if the source impedance is low. However, some transducers (piezoelectric transducers, for example) have high source impedances and should be used with a special charge-sensitive amplifier. In addition, multiplexing can greatly reduce a data acquisition system's effective input impedance. See Chapter 3 for detailed discussion.

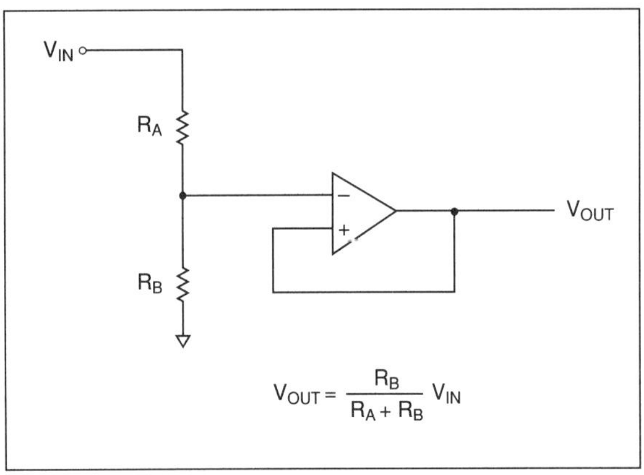

Fig. 6.07: Buffered attenuator

Attenuators & Multiplexing

There is a basic voltage range of 5V to 10V associated with most data acquisition inputs. Voltages over 10V are generally not readable without attenuation. Simple resistive dividers can easily attenuate a voltage, which is too high for a 5V input, but there are two drawbacks complicating this simple scheme. Voltage dividers have a substantially lower impedance to the source than direct analog inputs, and, they have very high output impedance to the multiplexer input. Consider this example, a 10:1 divider reading 50V. If a 900 KOhm and 100 KOhm resistor are chosen, to provide a reasonable 1 MOhm load to the source, the impedance seen by the analog input is a disagreeable 90.0 KOhm which will never allow an accurate multiplexed reading. If the values are both downsized by a factor of 100 to allow the output impedance to be lower than 1,000 Ohms, the input impedance seen by the measured source is 10 KOhm. Most voltage measurement applications could not tolerate a 2K Ohm/volt instrument. Hence the conclusion: simple attenuation is not practical with multiplexed inputs.

Attenuators & Buffers

The answer to the voltage divider problem is unity gain buffering of the voltage divider output. A dedicated unity gain buffer does not have a source impedance problem with 90 KOhms. The output impedance is very low to the multiplexed analog input as well.

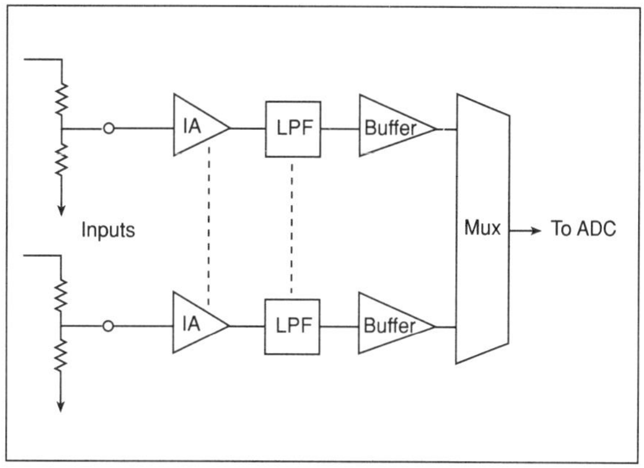

Fig. 6.08: Filtered/buffered data acquisition system

Low-Pass Filters

Low-pass filters attenuate higher frequencies to varying degrees depending on the number of stages and the magnitude of the high frequency relative to the corner frequency. There is no need for high bandwidth in an amplifier stage if the signal of interest is of a much lower frequency. In fact it is a basic intent of this design to eliminate excessive bandwidth in all circuits in order to limit noise. One of the major benefits of individual signal conditioning stages for low-level sensors, as opposed to multiplexed stages, is the opportunity to include low-pass filtering on a per-channel basis in the signal path. **Note:** In a multiplexed circuit (an amplifier which is being shared by multiple low level DC signals), the main signal path generally cannot low-pass filter due to the fast settling time necessary in multiplexed systems.

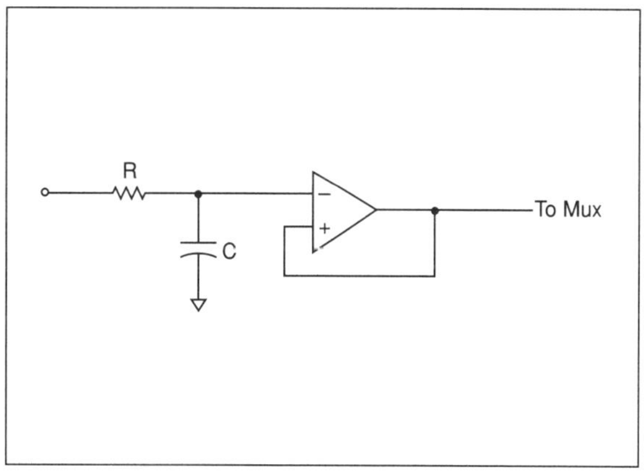

Fig. 6.09: Simple RC filter with buffer

The best place for low pass filtering is in the individual signal path prior to buffering and multiplexing. For very small signals, amplifying with an instrumentation amplifier prior to filtering allows an active low pass filter to operate at levels of optimum signal-to-noise ratio. Figure 6.10 on the following page illustrates a typical amplifier-filter-mux configuration.

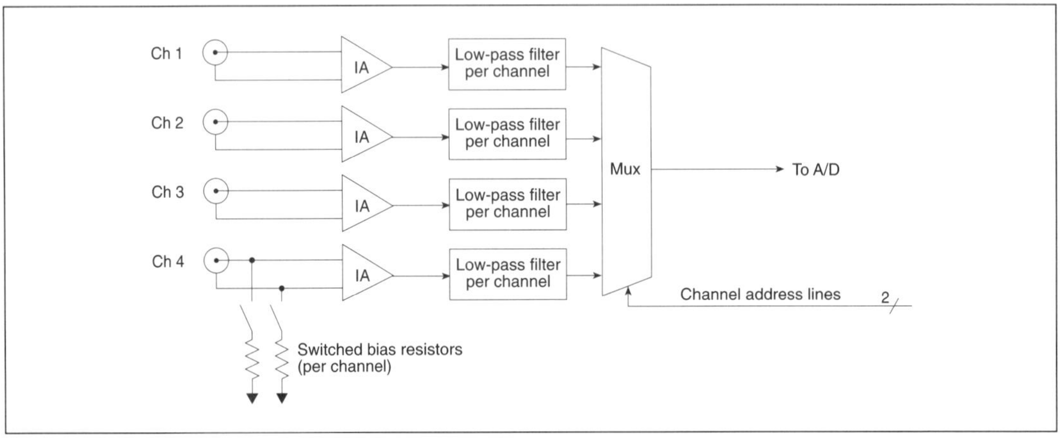

Fig. 6.10: Typical amplifier-filter-mux configuration

Measuring High Voltages

It was shown earlier that measurement of voltages above 5-10V with data acquisition inputs requires attenuation and that attenuation requires buffering. A typical data acquisition system would require a signal conditioning stage, which provides attenuation and buffering, in order to accommodate high-voltage inputs.

For higher voltage measurements a fully differential attenuator, calibrated to match the buffered differential inputs of a data acquisition system would be required. With this type of setup, differential voltages as high as 2000 volts peak-to-peak are measurable. Figure 6.11 illustrates this type of system.

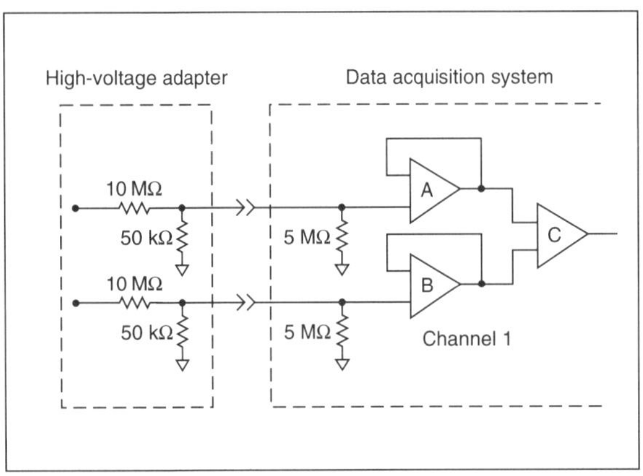

Fig. 6.11: Typical data acquisition system with a high-voltage adapter

For systems with a true differential input, neither the input's low or high end are connected to ground. In this case, attenuating an incoming signal requires two matched resistor dividers; one on signal-high and one on signal-low. The low end of each divider is grounded. For example, to provide 200:1 attenuation without compromising the high input impedance, an accessory device is required. In the case of our typical system, the two matched buffer amplifiers (2 op amps in the same package), A and B, on the front end of the data acquisition system draw nearly identical bias current through their inputs. If matched voltage dividers are placed on the high and low signal inputs, the bias current through the dividers will cause an identical voltage offset on the output of the op amps A and B. Since these two outputs are identical, they can be thought of as a common mode voltage that the differential amplifier, C, rejects. See Figure 6.11.

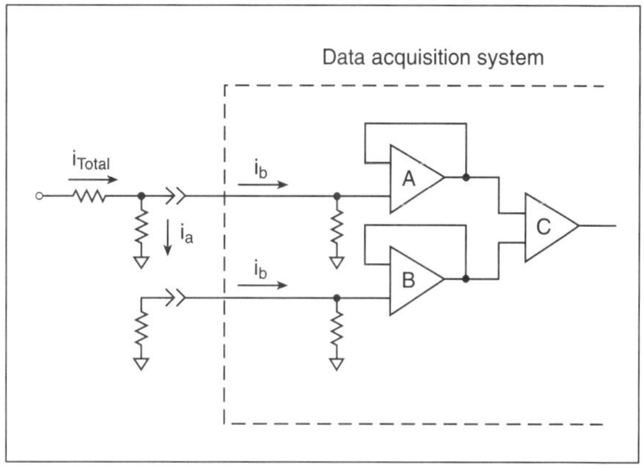

Fig. 6.12: Single ended attenuation of a differential input

This scheme attenuates both the high and low signal, allowing both ends to be hundreds of volts off ground. For example, the high input can be 110 VAC while the low end has a 50 VDC offset. After attenuation, the peak voltage of the high signal is 0.77V and the low signal is 0.25V, which is well within the readable range of the data acquisition system's differential inputs.

If a single ended measurement is attempted by attenuating only the high signal and pulling the low signal to ground, two problems are encountered. First, if the data acquisition system ground and the signal ground are dissimilar, tying them together creates a ground loop and produces error in the reading. The second error component results from the bias currents i_b flowing through the two op amps, A and B. The bias current produces a voltage drop through the divider on the high input, but there is no matching drop through the low input. The difference is presented to the differential amplifier, C, producing an offset on its output. See Figure 6.12.

Single-Ended & Differential Measurements

Data acquisition systems have provisions for single-ended and differential input connections. The essential difference between the two is the choice of the analog common connection. Single-ended multi-channel measurements require that all voltages be referenced to the same common node, which will result in measurement errors unless the common point is very carefully chosen; sometimes there is no acceptable common point. Differential connections cancel or ignore common mode voltages and allow measurement of the difference between the two connected points. When given the choice, a differential measurement is always better. The rejected common mode voltages can be steady DC levels or noise spikes.

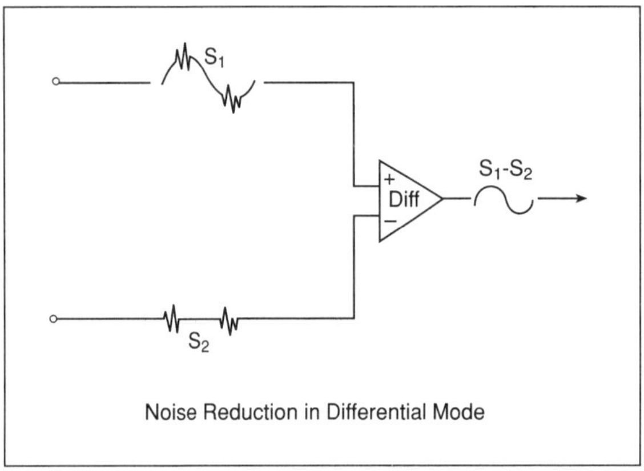

Fig. 6.13: Voltages common to both lines are rejected by the differential amplifier

The best reason to choose single-ended measurements is for a higher channel count that is available in some devices. Most data acquisition products allow you to double the number of channels in a differential system by selecting single ended operation.

There is a widely misunderstood aspect of differential measurements of floating sources. A common example is a flashlight battery connected to a differential input pair. The measurement cannot be made unless an additional connection is made to the analog common of the analog input device. This connection is frequently made with a 10K-100K resistor. The resistor provides a bias current return path for the two inputs which are each connected to high impedance voltage follower stages on the input of an instrumentation amplifier. If this bias current return path is not established, the amplifier will saturate and cannot measure the voltage difference between the inputs. Some data acquisition inputs provide switched resistors to analog common for this purpose. The reason these bias resistors are switched is that there are situations in which they are not needed. The situation in which the bias resistors are not needed is where a DC path already exists for bias current to flow from the amplifier inputs to analog common. This usually occurs when the device under test is already referenced to analog common.

Differential Voltage Measurements

In a single-ended voltage measurement, the voltage on the high input lead is measured with respect to analog common. Differential measurement utilizes a differential amplifier to measure the difference in voltage between the two signal leads. Differential measurements provide greater noise rejection; furthermore, some measurements, most notably strain-gage measurements, require measurement of the voltage difference between two signal leads.

Figure 6.14 depicts a differential amplifier configured for a thermocouple measurement. Note the 10 KOhm resistor between the low side and ground. Even though the amplifier measures the voltage difference between the two inputs, a path to ground is required on at least one input of the amplifier. This resistor provides the path for the amplifier bias currents. If not present or too high in value, the amplifier bias currents will cause the inputs to approach one of the power supply rails and exceed the common mode range of the amplifier. In most applications, the common mode range is typically ±10V. The common-mode voltage produced by the bias currents is $V_{cm}=2i_bR_b$

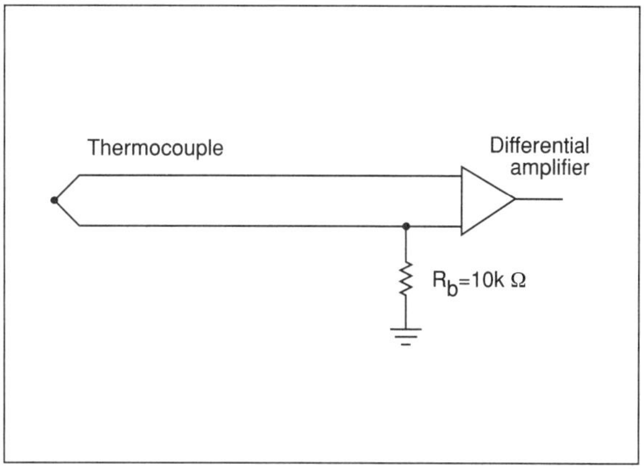

Fig. 6.14: Differential thermocouple measurement

To determine the minimum value for R in the case of a completely floating input, R may be a short circuit on one side only. Otherwise, under the worst case conditions, the voltage error in percent of voltage applied due to source resistance and R is: $\Delta E(\%)=100\ R_s/2R_b$. This applies to R for the balanced configurations. Typical values for R are 10 KOhm to 10 MOhm. If both equations cannot be satisfied, the source resistance is too high to operate within specifications.

Some amplifier inputs have built-in termination resistors for differential measurements, but very high impedance inputs require the user to supply a termination resistor. Introducing a termination resistor into the configuration dramatically lowers the input impedance in non-isolated systems and may load the circuits being measured.

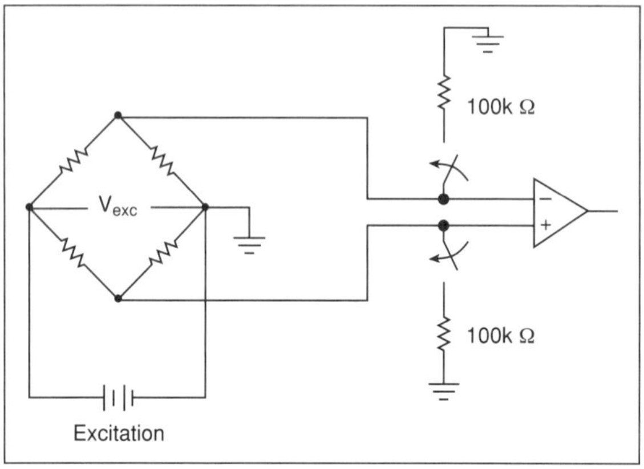

Fig. 6.15: Differential strain gage measurement with switchable termination resistors

Consider the strain gage in Figure 6.15. If the excitation voltage is referenced to the same common as the differential amplifier, termination resistors are not required because the strain gage itself provides a current path to ground for the instrumentation amplifier. However, if the excitation voltage is floating or referenced to a separate common, at least one termination resistor is required. A balanced termination with 100 KOhm resistance can also be used.

Common mode rejection ratio (CMRR) is an important specification for instruments making differential measurements. A large CMRR is necessary to accurately measure a small differential signal on a large common mode voltage. The CMRR specifies the maximum effect of the common mode voltage on a differential measurement.

An amplifier's CMRR is typically specified in decibels (dB); a ratio of 10 results in 20 dB, and each additional factor of 10 results in an additional 20 dB. For example, 100 dB corresponds to 10^5, so an instrument with a CMRR of 100 dB measuring a signal on a common mode voltage of 1V can have an error as large as 0.01 mV. If the common mode voltage is larger, or if greater accuracy is desired, an amplifier with a CMRR of at least 120 dB would be required. See Chapter 5 for detailed discussion.

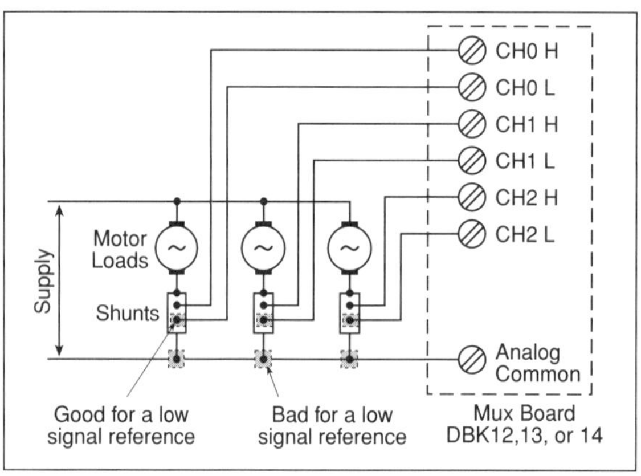

Fig. 6.16: Ground referenced differential measurement

Measuring High DC Currents with Shunts

It is frequently necessary to measure high currents with data acquisition instruments. Generally, DC currents are measured by monitoring the voltage drop across a calibrated shunt which has a 50 mV or 100 mV drop at full rated current. Figure 6.16 shows a differential measurement of three shunts which have been strategically placed in the common sides of three motor circuits to allow inexpensive monitoring with a non-isolated, low voltage input. It is not always possible to locate the shunts in higher voltage circuits in the electrically convenient common leg. In these instances, isolated analog inputs are the recommended solution.

Measuring AC Currents with Current Transformers

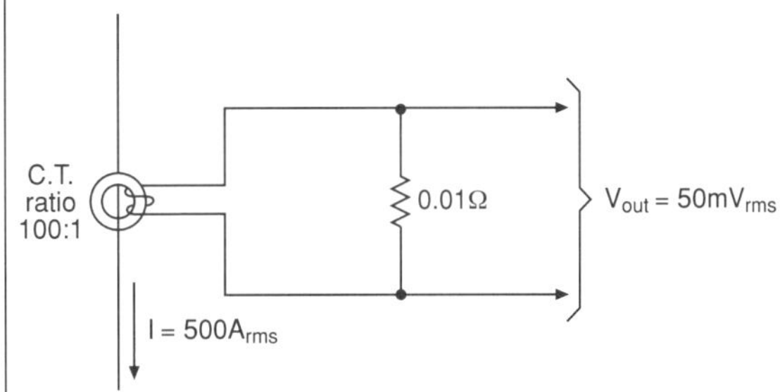

Fig. 6.17: Measuring high AC currents with current transformers

High AC currents can be measured with shunts also, but the inherent danger of direct connection to AC lines is undesirable. Isolation from the AC line voltages is provided by current transformers (CT) along with a ratio reduction of the input current. A 500:5 CT for example, has a 100:1 ratio and will have a secondary current of 5 amperes when 500 amperes is flowing in the primary. If a low value load resistor such as 0.01 Ohm is connected to the output of a current transformer, the full load secondary current will produce a 50 mVACrms, which can be easily read with an analog input. Although the output voltage seems low, higher resistance values will overload the CT and reduce the accuracy of the measurement. If a CT has a VA rating of 2 VA, the maximum usable load resistance including lead wire is only 0.08 Ohms. It is also necessary to remember that an open-circuited CT with primary current flowing can generate over 1,000 volts at lethal current levels.

Signal Connection Methods

Measurement of low level signals depends heavily on proper wiring between the signal sources and the data acquisition device. The amplifier stages in signal conditioning equipment are not able to determine the difference between the signal and inadvertently added noise voltage due to improper wiring practices. Whenever low signal levels (less than 1V) are being measured, twisted pairs or shielded twisted pairs provide the best protection against unintentional pickup of noise voltages. If a shield is used, it should only be grounded at one end, preferably at the signal source.

Chapter 7
Noise Reduction & Isolation

Noise Reduction

Controlling electrical noise is imperative because it can present problems even with the best measurement equipment. Most laboratory and industrial environments suffer from multiple sources of electrical noise. For example, AC power lines, heavy equipment (particularly if turned on and off frequently), local radio stations, and electronic equipment can create noise in a multitude of frequency ranges. Local radio stations are a source of high-frequency noise, while computers and other electronic equipment can create noise in all frequency ranges. Creating a completely noise-free environment for test and measurement is seldom practical. Fortunately, simple techniques such as using shielded/twisted pair wires, averaging, filtering, and differential voltage measurement are available for controlling the noise in our measurements. Some techniques prevent noise from entering the system; other techniques remove noise from the signal.

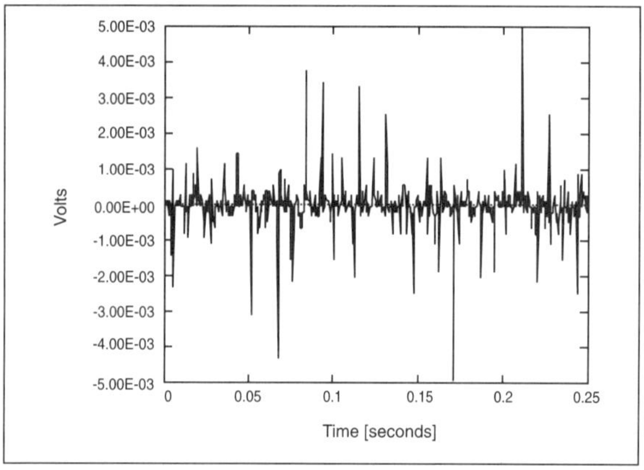

Fig. 7.01: Single-ended measurement with loose wires

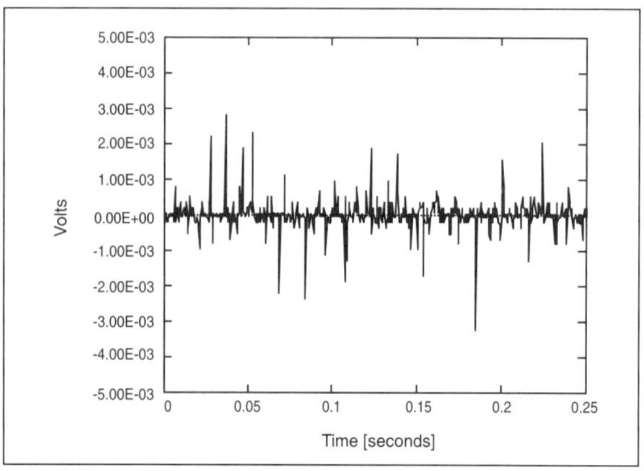

Fig. 7.02: Single-ended measurement with shielded wire

Shielding. While many techniques for controlling noise in signals provide a means of removing noise that is already present, the preferred solution is simply preventing the occurrence of noise in the signal in the first place. This is accomplished through careful shielding of equipment and test leads.

For example, loose wires, which are effective antennae for radio frequency pickup, and can form loops for inductive pickup, are a source of noise; using shielded cables is a simple solution.

To illustrate the need for controlling noise, Figure 7.01 shows a single-ended voltage measurement on a shorted channel. Approximately 6 feet of wire, which was not shielded or twisted pair, was attached to a data acquisition system.

Figure 7.02 shows the noise in the measurement of a single-ended shorted channel, where shielded cable is the noise reduction technique used. The improvement over unshielded wires is readily apparent.

Averaging. Averaging is done in software after several samples have been collected. Depending on the nature of the noise and the technique used, averaging can reduce noise by the square root of the number of averaged samples, however, reducing RMS noise significantly may require averaging many samples. Figure 7.03 demonstrates the voltage across the shorted channel when 16 samples of data are averaged together. Averaging is best suited to lower-speed applications.

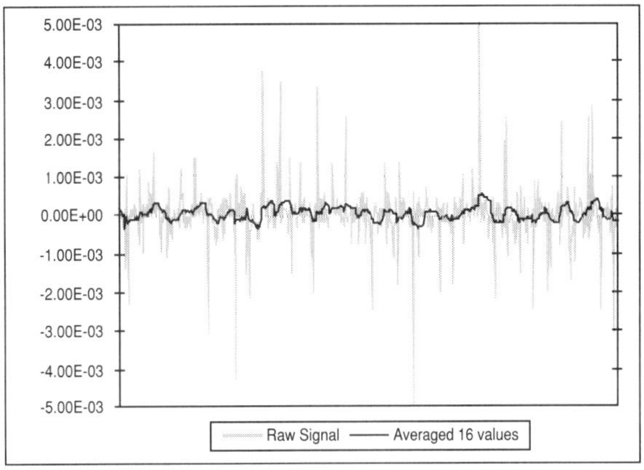

Fig. 7.03: Noise reduction using averaging

Although averaging is an effective technique, it suffers from several drawbacks. Noise present in measurements only decreases as the square root of the number of measurements. Therefore, in the above example, reducing the RMS noise to a single count by averaging alone would require 3,500 samples. As such, averaging is only suited to low-speed applications. Also, averaging only eliminates random noise. It does not necessarily eliminate many types of system noise—such as periodic noise from switching power supplies.

Analog Filtering. A filter is an analog-circuit element that attenuates an incoming signal according to its frequency. Depending on whether or not the filter is a low- or high-pass filter determines if the frequencies are attenuated above or below the cutoff frequency. For example, as a signal's frequency increases beyond the cutoff point, the attenuation of a single-pole, low-pass filter increases slowly. Multi-pole filters provide greater attenuation beyond the cutoff frequency but may introduce phasing problems that could affect some applications. Filter circuits can be passive or active.

Passive. A passive filter is a circuit or device, consisting entirely of non-amplifying components, typically inductors and capacitors which pass one frequency band while rejecting others.

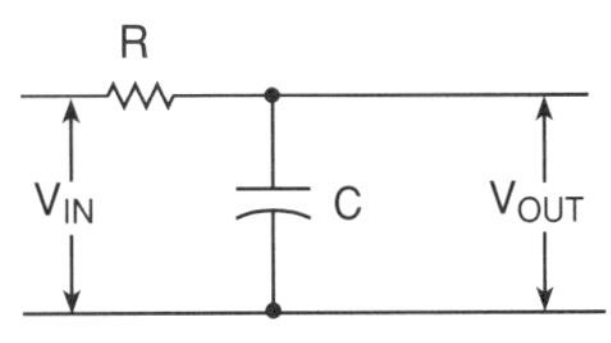

$$F_{-3dB} = \frac{1}{2\pi RC}$$

Fig. 7.04: Single-pole RC filter with the formula for -3dB frequency

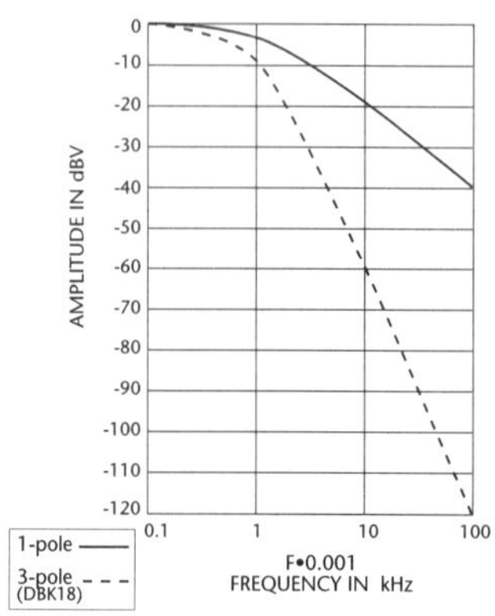

Fig. 7.05: Amplitude vs. frequency for a typical single-pole low-pass filter vs a three-pole filter

Active. An active filter is a circuit or device, made up of amplifying components and suitable tuning elements, typically resistors and capacitors, which pass one frequency band while rejecting others. Figure 7.05 compares the amplitude of a single-pole, low-pass filter, with a three-pole filter. Both filters are set for a 1-kHz cutoff frequency. The three-pole filter has a much greater attenuation for frequencies exceeding the cutoff frequency. The improvement in signal quality provided by low-pass filtering is demonstrated in Figure 7.06, in which a signal containing wideband noise is passed through a three-pole filter configured for a 1-kHz cutoff frequency. The deviation from the average signal is plotted in volts. The maximum deviation is 6 counts, and the RMS noise is 2.1 counts.

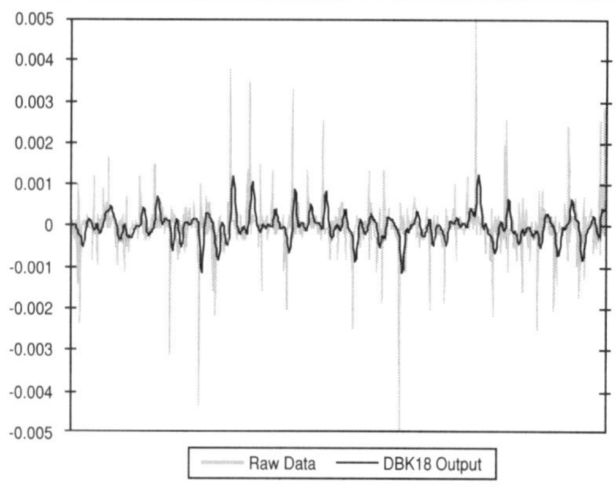

Fig. 7.06: Noise using a typical low-pass filter card

The three-pole filter used in our example is an active filter input with changeable filter configurations. The active 3-pole filter can be configured as a Butterworth, Bessel, or Chebyshev filter with corner frequencies up to 50 kHz. Filter properties depend on the values of the resistors and capacitors, which can be changed by the user. Filtering can also be done with switched-capacitor filters. This type of filter requires a clock signal to set the cutoff frequency. The primary advantage to this type of filtering is the ease of cutoff frequency programmability.

Differential Voltage Measurement. Making differential voltage measurements is another means of reducing noise in analog input signals. This technique works because often most of the noise on the high analog input lead closely approximates the noise on the low lead. This is called common mode noise. Measuring the voltage difference between the two leads eliminates common mode noise.

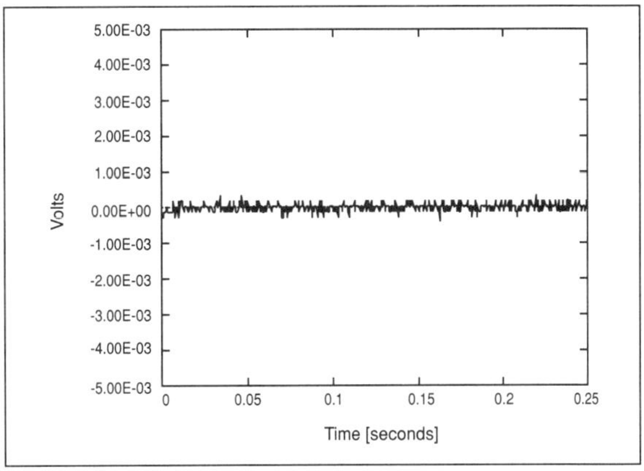

Fig. 7.07: The same signal as Figure 7.01 using a differential input

The improvement obtainable with differential voltage measurement is illustrated in Figure 7.07, which shows the same signal as Figure 7.01 using a differential input rather than single-ended.

Isolation

Isolation can be defined as the separation of one signal from another to prevent unintentional interaction between the two signals. There is a degree of channel-to-channel isolation in any multiplexed data acquisition system; relay-based systems have galvanic isolation and solid state systems do not. Galvanic isolation is the absence of any direct current (DC) path. Most isolation methods eliminate any DC path below 100 MOhm. Three benefits of galvanic isolation are circuit protection, noise reduction, and rejection of high common-mode voltages, especially those created by ground loops.

Circuit Protection. Isolation separates the signal source from measurement circuitry that may be damaged by the signal. Voltages higher than about 10V can distort data or damage components used in data acquisition systems. High-voltage signals or signals with high-voltage spikes should therefore be isolated. The protection can also work in the other direction—to safeguard a sensitive signal conditioner from a device failing elsewhere in the system.

Noise Reduction. Isolation eliminates ground loops for high-gain systems and multi-unit systems that are grounded together. The chassis for each device can reside at a ground potential slightly different from that of the other devices. These potential differences cause the flow of ground currents that result in measurement errors.

Rejection of High-common Mode Voltage. Common-mode voltage, relative to earth ground, is present in most measurement situations. Two voltage measuring points which are relatively close together in potential, may share a very high common-mode voltage. In order to make a safe and accurate measurement in this situation, all data acquisition systems must use some method of rejecting common-mode voltage. The instrumentation amplifier in a simple MUX-IA-ADC data acquisition system is only capable of rejecting about 10V of common-mode voltage. Common-mode voltages higher than 10V require the use of more complicated and expensive rejection methods, such as isolation amplifiers, which employ magnetic, optical, or capacitive isolation.

Methods of Isolation

There are many methods to achieve isolation, with the three most common being magnetic, optical and capacitive.

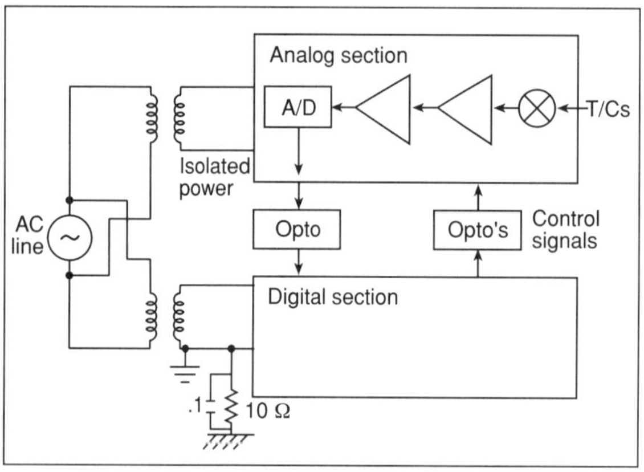

Fig. 7.08: Block diagram of a typical isolation scheme

Magnetic Isolation. Magnetic isolation using a transformer, is the most common method to isolate analog signals, and is also capable of isolating digital signals, although not as common.

The power supplies in most systems use transformers to isolate the AC line voltage from the DC voltages generated to power the data acquisition system. A transformer provides isolation by magnetically coupling the AC input power to the data acquisition system, thereby providing no DC path for current to flow.

Transformer isolation is also commonly used in input amplifiers to provide isolation from the input signal to the measurement circuitry. In this application, the input signal is first converted to an AC signal which can be fed to the primary of the transformer. The AC signal is either frequency, pulse width, or amplitude modulated, and then magnetically coupled to the secondary winding. The output is then demodulated by the receiving circuitry before being presented to the measurement circuitry.

Finally, transformers can also be used to isolate digital signals. This method is increasingly rare as the cost and performance of optical isolators improves (see next discussion).

Optical Isolation. Optical isolation is most commonly used to transmit digital signals. A light-emitting diode within an opto isolator generates light when current flows through the diode. A receiving transistor, which is light sensitive, will enable a current to flow when it detects light. Since the only connection between the input and output is a path for light to pass, then the device can provide isolation of several thousand volts from input to output.

Opto-isolators are very commonly used to isolate the output of an A/D converter. The A/D's output is usually serial (or parallel which has been converted to serial) to minimize the number of opto's required in a system (parallel to serial conversion is less expensive than having one opto for each bit of output from the A/D). In these instances, the power supply to the A/D and associated input circuitry is also isolated, usually via a transformer as described above.

Capacitive Isolation. A capacitor is a passive electronic device that permits AC current to flow through it, but will not pass a DC current. As a result, it can be an inexpensive and simple device for isolation. In order to be used, the signal which is to be isolated must first be converted to an AC signal through some modulation technique, and then transmitted through the capacitor to the receiving side. At the receiving side, the AC signal must again be demodulated in order to recreate the signal to be measured. This technique is applied within low cost isolation amplifiers in which the coupling capacitor is formed by a common layer between two isolated IC substrate sections. Signal isolation equipment using these specialized IC's can achieve isolation ratings as high as 1500 VAC.

The main benefits of this approach are simplicity, low cost and bandwidths as high as 50 kHz. Figure 7.09 illustrates a DC/DC converter often used as a component subsection in an isolation amplifier.

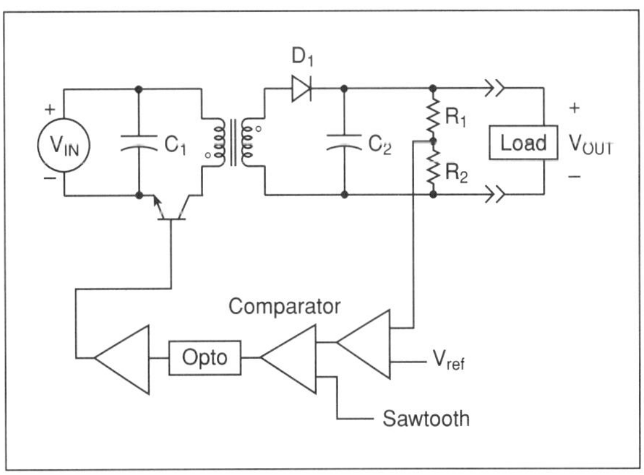

Fig. 7.09: Diagram of DC-to-DC converter

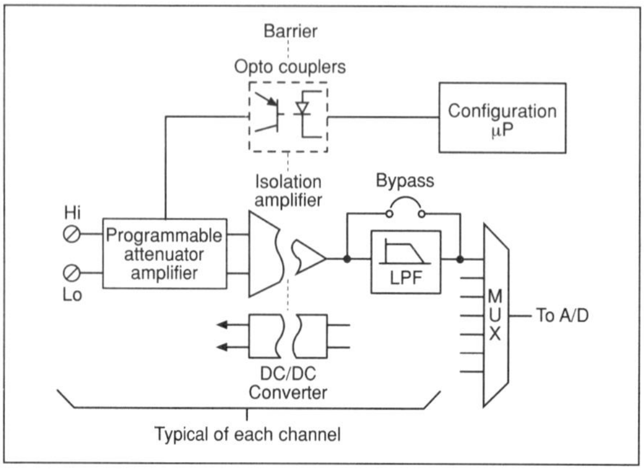

Fig. 7.10: Multichannel programmable isolated amplifier block diagram

Figure 7.10 illustrates a typical multichannel programmable isolation amplifier, in a data acquisition system that uses three types of isolation: transformer, optical, and capacitive. A transformer-based DC-to-DC converter is used to supply power to the isolated side. A capacitively coupled device is used to isolate the analog signal, while two opto-couplers are used to transmit the digital control signals to the floating circuitry.

IOtech's MultiScan/1200 high-speed isolated temperature and voltage measurement instrument, which provides a comprehensive isolation scheme, uses a combination of relays and transformers to provide 200V of isolation on a per channel basis. See Figure 7.11. Adjacent thermocouple channels can have signals that are dissimilar by 200V and still read temperatures accurately.

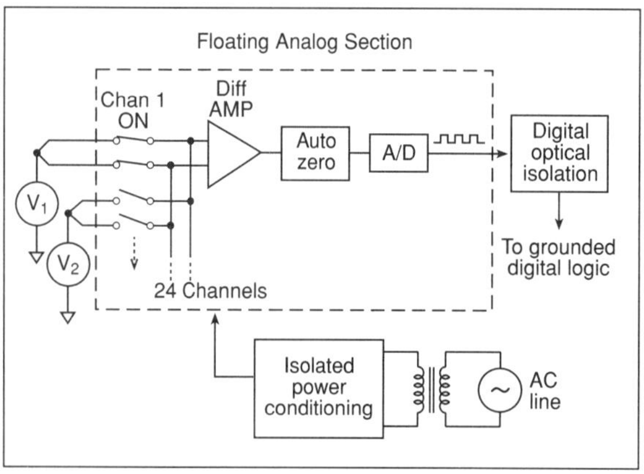

Fig. 7.11: Analog input block diagram

To provide this capability, the MultiScan/1200's analog front end, including its A/D converter, is isolated from system ground via the AC power supply's transformer. The digital information to and from the analog section is transmitted via opto-couplers. Each MultiScan/1200 analog input card has a pair of relays on each channel. At any single moment, only one input channel is attached to the analog path. As each channel is switched in, the entire analog front end floats to the ground reference of the newly connected channel.

Chapter 8
Digital & Pulse-Train Signal Conditioning

Digital Signals

Many applications entail interfacing digital signals to a computer. Digital signals differ from analog signals in that digital signals represent a discrete state, whereas analog signals represent continuous values. Digital signals are termed discrete because a digital line is either low or high. As such, a digital signal represents Boolean values such as off or on, zero or one. Accordingly, each digital line can serve as a single bit in the representation of a number or character. Digital output lines commonly control relays to power-on or power-off equipment or indicators. Digital input lines can represent the state of a switch indicating position or status. Digital I/O lines may also be used to communicate between instruments.

Relays. A relay is an electromechanical switch. Most relays open and close a switch using a magnetic field produced when current is passed through a coil. Consequently, the relay behaves like an isolator in relation to the control current. A relay will be either open or closed when control current is not flowing. Most relays specify the control current and voltage required to operate the switch, as well as the current and voltage ratings for the power transmitted through the switch.

Relays are typically used to "break" or "make" circuit paths that carry high voltage or current. A relatively small current supplied to the magnetic coil can throw a switch capable of carrying hundreds of amps.

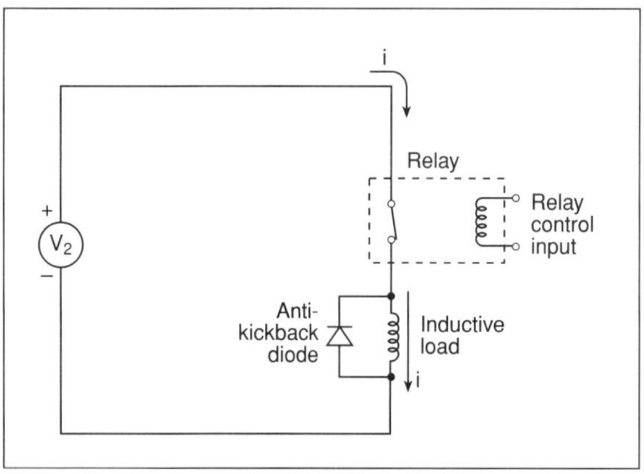

Fig. 8.01: Relay turning power on and off to an inductive load

Relays are commonly controlled by a digital output from a computer or other instrument. For example, to control a heater, an instrument monitoring temperature can use its digital output to actuate a high-voltage relay attached to the heater coil.

When an inductive load is being switched, a diode across the load provides a path for the current spike directly after the switch is opened while the inductor's magnetic field is collapsing. Without the diode, arcing will occur at the relay's contacts, potentially decreasing the life of the relay. See Figure 8.01.

Relays can also provide isolation in a data acquisition system. For applications where each channel must be allowed to float, or where individual channels have high common mode voltages, relay isolation is an economical and effective means of isolation.

Because they are an electromechanical device, relay contacts can "bounce" as they are closed. As the coil is energized, the contacts close abruptly, and for a brief moment, the contacts can rattle into place causing momentary interruptions in the current. Bounce is typically not a problem when switching loads on and off. Data acquisition devices utilizing relays generally provide debounce time to ensure that the relay has fully settled before initiating a reading.

High Current/Voltage Digital I/O. Digital circuits are not limited to low-level TTL and CMOS systems. For some applications, the high level may typically be 30V and the source current may commonly be 100 mA.

In applications where a relay coil must be energized, a TTL or CMOS output is rarely adequate. To provide the necessary current and voltage, a buffer stage must be added between the TTL signal and the relay coil. One example of this type of system is IOtech's Digital488/80A digital I/O instrument. This type of instrument can be used to switch a 24V relay on and off as long as an amplifier stage is provided. In the case of the Digital488/80A, the amplifier/attenuator, consisting of a PNP transistor, a flyback diode, and a resistor, is provided by an option card, which can be added internally to the unit.

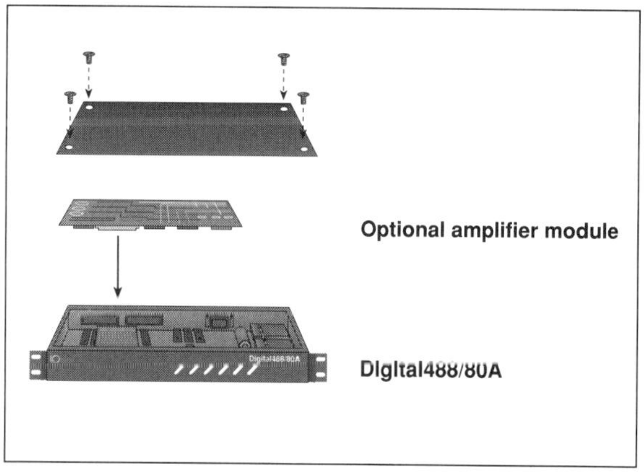

Fig. 8.02: This illustrates the optional amplifier module that installs inside the Digital488/80A to accommodate high voltages and/or current

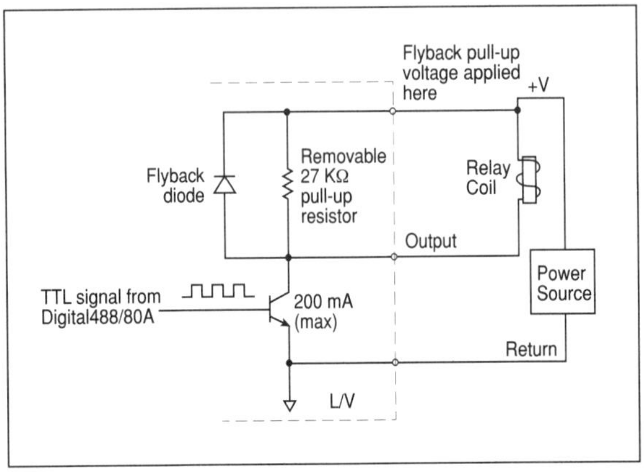

Fig. 8.03: A typical digital-output signal amplifier similar to that used in IOtech's HVCX1

To energize a 24V solenoid or relay using this type of instrument, an external 24V supply is attached to the circuit. As the internal TTL output goes high, the transistor is biased and the output goes low (about 0.7V). When the TTL output is low, the transistor does not conduct and the output goes to 24V. Since the relay coil represents an inductive load, the flyback diode should be attached so that the circuit is not damaged during switching.

Using the IOtech's Digital488/80A as an example, Figure 8.04 shows high-voltage digital inputs via the HVCX1 attenuator circuit. This allows voltages up to 48V to be read by the TTL circuitry.

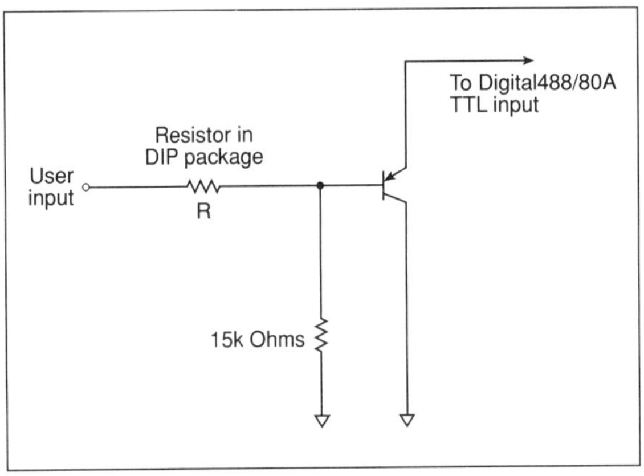

Fig. 8.04: Schematic of typical attenuator input, similar to that used in IOtech's HVCX1

The high-voltage signal goes into a resistor voltage divider, which acts as a signal attenuator. The resistance value, R, provides a means for selecting the Boolean "high" voltage level. The table below shows the resistor values for frequently used voltage levels.

Typical Voltage Levels & Associated Resistor Values

Inputs	Resistor Value
0-5V	10 Ω
0-12V	20K Ω
0-24V	56K Ω
0-48V	120K Ω

Speed, Timing, and Scanning of Digital Inputs

There are many ways to interface digital inputs to a computer, ranging from simple to complex. In this section we will briefly discuss software-triggered single-byte readings, hardware-paced digital-input readings, and externally triggered digital-input readings.

Asynchronous Digital-Input Readings. In cases where the computer needs to periodically sample a digital byte or a group of bits, a software-triggered, asynchronous read is required. Sometimes, though, the speed and timing of the digital-input readings are important. The time between readings is likely to vary when using the software-triggered single-byte method, particularly in applications implemented under a multitasking operating system. This is because the time between readings depends on the speed of the computer and other tasks that must be performed concurrently. Variations in time between readings can be partially compensated for by the use of software timers, but it is difficult to obtain a timing resolution of better than 1 ms (millisecond) on a PC.

Synchronous Digital-Input Readings. Some systems, such as IOtech's DaqBook® and WaveBook/512 portable data acquisition systems, offer hardware-paced digital-input readings. In such systems, the digital input port can be read at a preset frequency. For example, the DaqBook/200 offers a 16-bit port that can be read at up to 100 kHz, and the WaveBook/512 has an 8-bit port that can be read at up to 1 MHz. The greatest advantage of hardware-paced digital-input readings is that they can be implemented far more quickly than can software-triggered readings. In addition, the time between such readings is constant. Finally, devices such as the DaqBooks and WaveBook/512 are able to place the digital-input port reading into the scan group of their analog readings, providing a close correlation between analog and digital input readings.

Externally Triggered Digital-Input Readings. Some applications use external devices that provide digital data at a pace independent of the data acquisition system. These devices provide a digital bit, byte, or word that needs to be read by the data acquisition system. It is often necessary to take readings only when new data is available rather than at a predetermined interval. Because of this, such external devices typically have handshaking ability. As new digital information becomes available, the external device issues a digital transition on a separate line such as an External Data Ready or Strobe input. To interface with a device like this, the data acquisition system must provide an input latch, which can be controlled by this separate external signal. Furthermore, logic must be supplied to the controlling computer, alerting it to the fact that new data is ready to be received from the latch.

One example of a device that operates in this fashion is IOtech's Digital488 series of interface products. The Digital488 has an Inhibit line, among its six handshake/control lines, for notifying external devices that the input latch is being read. This allows the external device to hold new digital information until the current read is successfully performed.

Digital Isolation

Digital signals are isolated for several reasons:

- to protect each side of the system from an inadvertent overvoltage condition on the other side
- to facilitate communication between devices with different grounds
- to prevent injury when circuits are attached to individuals in medical applications

Optical Isolation. Optical isolation involves the use of an LED or diode laser to transmit the digital signal and a photodiode or phototransistor to receive it. This technique is depicted in Figure 8.05. IC-sized opto-couplers provide voltage isolation in the range of 500V. Using this technique, digital devices with dissimilar grounds can be controlled and monitored.

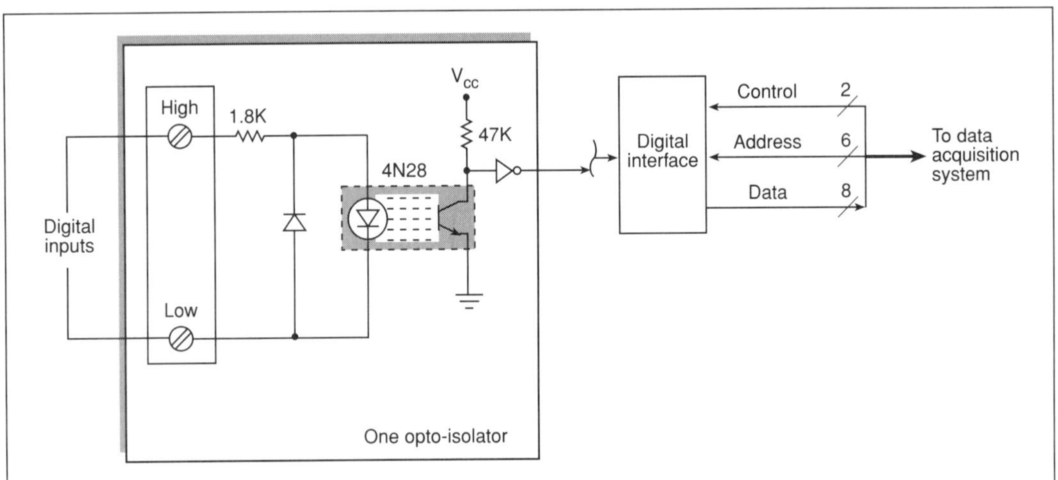

Fig. 8.05: Schematic of opto-coupler isolating a system with its own ground

Pulse-Train Signal Conditioning

In many frequency measurement applications, pulses are counted and compared against a fixed timebase. A pulse can be thought of as digital rather than analog since only the number of rising or falling edges are of interest. In many instances, however, the pulse-train signal comes from an analog source, like a magnetic pick up.

IOtech's DBK7 frequency-input card provides the DaqBoard, DaqBook, and Daq PCM-CIA data acquisition systems with four-channels of frequency input via two separate front-end circuits, one for true digital inputs and one for analog inputs. An analog input is a signal that is time-varying but not digital by nature.

The DBK7 conditions digital inputs of different levels in the same manner as the Digital488/80A's amplifier module discussed earlier. The analog input circuit is used to convert a time-varying signal into a clean digital pulse-train.

Figure 8.06 shows the schematic of the analog-input signal-conditioning path. The front-end RC network provides AC coupling, allowing all signals above 10 Hz to pass. The selectable attenuator desensitizes the circuit from unwanted low level noise by reducing the overall magnitude of the waveform.

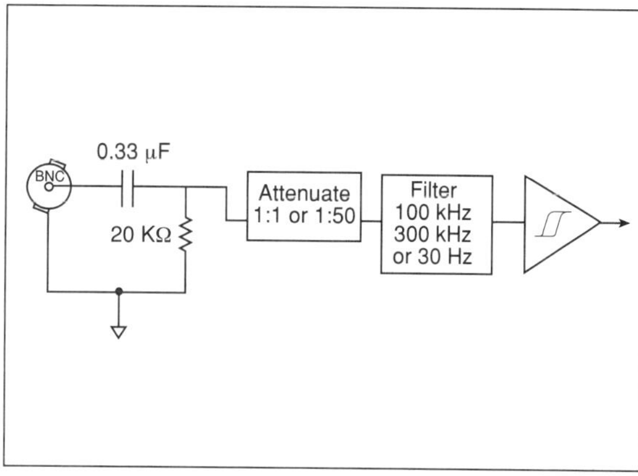

Fig. 8.06: The DBK7 front end for converting analog signals to a digital pulse stream

When using a pulse-train from a relay closure, the DBK7 provides programmable debounce settings that allow the user to select the amount of debounce time required. The digital circuitry performs debouncing by monitoring the conditioned pulse-train for a sustained "low" or "high" level. If debouncing is not performed, the extra edges in the signal will produce an erroneously high, erratic frequency reading. Se Figure 8.07.

Many transducers produce an output whose information content is not in the signal's amplitude but rather in the signal's frequency. Sensors that measure rotational motion and flow typically fall into this class. Photomultiplier tubes and charged-particle detectors can also be used for measurements that require pulse counting. In principle, such signals could be sampled with an ADC, but this results in much more data than is necessary and makes the analysis cumbersome. Direct frequency measurement is far more efficient.

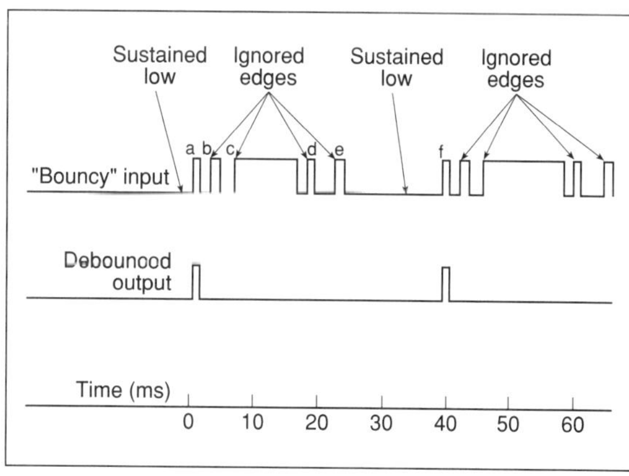

Fig. 8.07: The DBK7 debounce feature prevents relay bounce edges from being counted

Frequency-to-Voltage Conversion

Frequency-to-voltage conversion is commonly used to measure an input frequency via an ADC. The frequency-to-voltage converter's output is a DC-level signal whose magnitude is proportional to the input frequency. One of two methods can be employed to convert a pulse-train or AC signal into a DC level: pulse-train integration and digital-pulse counting.

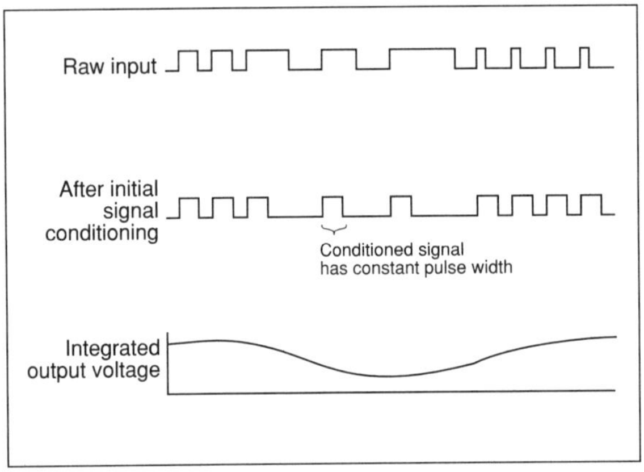

Fig. 8.08: Frequency-to-voltage using pulse integration

Pulse-Train Integration. For single-channel modular signal conditioning, 5B modules can be employed. The 5B frequency-to-voltage module is a good example of a pulse integrating F/V (frequency-to-voltage). A series capacitor provides AC coupling, removing any very-low frequency or DC signal. A comparator circuit produces a pulse with a constant duration each time the input signal passes through zero. The resulting pulse is then passed through an integrating circuit (a low-pass filter), producing a slowly changing level proportional to the input frequency, as shown in Figure 8.08.

The response time of the frequency-to-voltage converter is slow; it is the inverse of the cutoff frequency of the low-pass filter. The cutoff frequency of the low-pass filter should be much lower than the input frequencies of interest, yet high enough to provide the required response time. As the measured frequency decreases toward the cutoff frequency, significant ripple in the output becomes a problem, as shown in Figure 8.09.

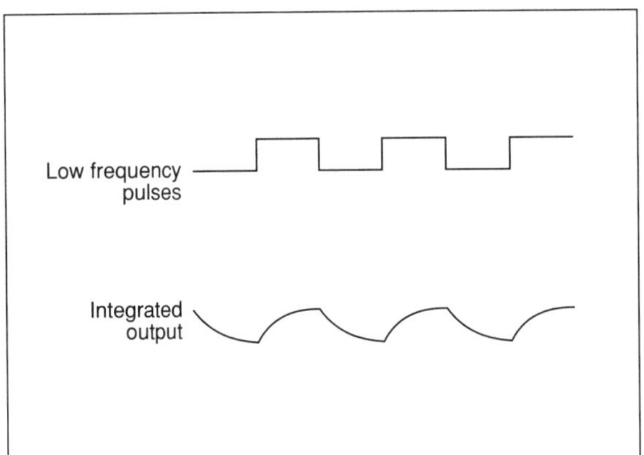

Fig. 8.09: Output ripple seen at low frequencies

Integrated circuits (ICs) that provide frequency-to-voltage conversion are available. Because an external capacitor is used to select the time constant, one can use a single IC to measure signals in vastly different fixed frequency ranges. To change range, the capacitor would have to be changed.

Frequency-to-voltage conversion works poorly for frequencies below 100 Hz. This is because a low-pass filter with cutoff frequency of under 10 Hz requires an extremely large capacitor.

Digital-Pulse Counting. Digital-pulse counting can be used to ensure a DC voltage that is proportional to the input frequency. IOtech's DBK7, frequency-input card uses digital-pulse counting in this manner. Front-end circuitry converts the incoming analog or digital signal into a clean pulse-train, devoid of relay contact bounce, high frequency noise, and other unwanted artifacts. A microcontroller is then used to accurately measure a total period consisting of several cycles extending over one user-selectable minimal period; this minimal period determines the frequency resolution. The microcontroller computes the frequency from the measured period and converts the frequency to a command that it issues to a D/A converter, which provides the DC level to the data acquisition system

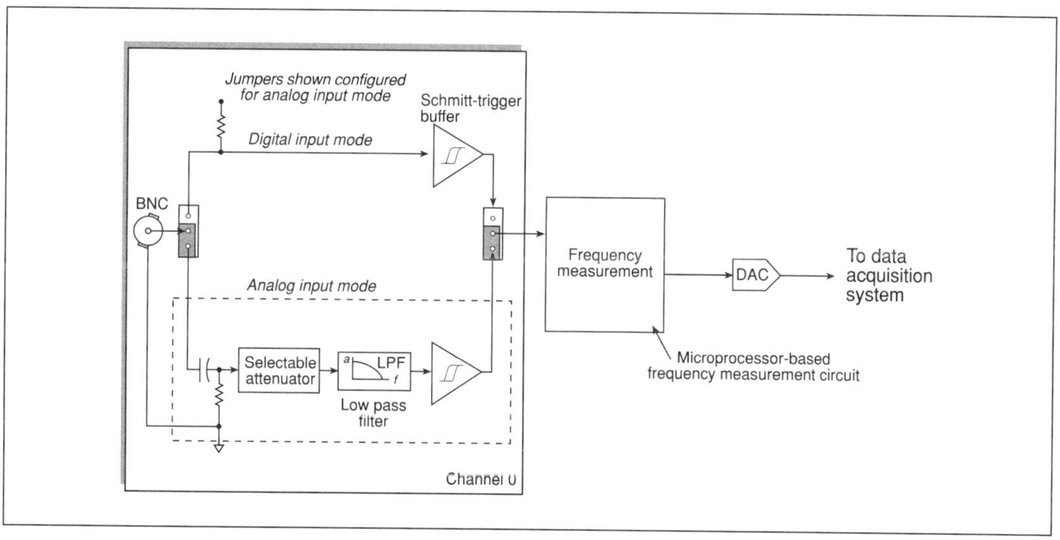

Fig. 8.10: DBK7 frequency-input card, signal path

The advantages of this method are that very low frequencies can be measured, the frequency range can be very wide, the output update can be quite fast, and the frequency range can be programmed, allowing the expected frequencies to use the entire analog-to-digital converter's range.

The output range of the digital-to-analog converter is +5V to -5V divided into 4,096 parts. The minimum frequency selected by the user becomes the -5V output, while the maximum frequency becomes +5V. The output update rate is regulated by F_{min} and F_{max} according to the following equation:

$$\text{Minimum measurement period (sec)} = (4{,}096 \times 0.5\ \mu S) \times (F_{max}/(F_{max} - F_{min}))$$

For example, for a system with F_{max} = 7 kHz and F_{min} = 500 Hz, the digital-to-analog converter output would be updated every 2.2 ms, or at a rate of 453 Hz.

In this equation, 4,096 is derived from the 12-bit digital-to-analog converter used, 0.5 µS is the resolution of the DBK7's timing circuits, and $F_{max}/(F_{max} - F_{min})$ is the ratio of the measurement time that must be increased to achieve 12-bit accuracy over the selected range.

Frequency Measurement by Gated Pulse Counting. Gated pulse counting can be used to measure frequencies much more accurately than can be done with frequency-to-voltage conversion. Gated pulse counting entails counting pulses that occur over a specified period of time. One determines the frequency by dividing the number of pulses by the counting interval. The error can be as low as the inverse of the counting interval. (For example, if the counting interval is two seconds, the error can be as low as 0.5 Hz.)

Many data acquisition systems include TTL-compatible counter/timer ICs that can be used to perform gated pulse digital level input. As such, they are unsuitable for use directly with unconditioned analog signals. Fortunately, many frequency-output devices feature a TTL output option.

IOtech's DaqBook and DaqBoard family of products use the 9513 counter/timer, which features five counter/timers. Many counter/timer ICs, including the 9513, generally use an oscillator built into the data acquisition system, or an external oscillator. Such ICs usually have several channels available to facilitate counting applications. Each channel features an input, a gate, and an output. The simplest counting method only uses the input. In this configuration, the PC is programmed to periodically read and reset the counter.

This approach's weakness is uncertainty in the timing interval. Variations occur in the execution speeds of the functions that begin and end counting. In addition, the function call that delays execution of the program for 50 ms employs an inaccurate software timer. These two effects can render a short-counting interval-frequency measurement useless. However, the technique is probably sufficient for counting intervals of greater than one second.

Gating can achieve greater accuracy because the gate controls the counting interval. As a result, frequency measurements are independent of any software timing issues. The gate can be configured so that pulses are counted only when a high signal enters the gate. Similarly, the gate can begin counting when it detects one pulse and can stop counting when it detects another.

A disadvantage of gated pulse counting is that it requires an extra counter to provide the gate. However, in multiple-channel applications, a single counter can provide the gate for many channels. For example, in a five-channel system, four channels count while one channel provides a gate.

Timing Applications. You can also use a data acquisition system with a counter/timer in timing applications. This can be accomplished by connecting a clock signal to the input of a channel and using the input signal as a gate. This method requires that the gate be configured for counting when the gate input is high. A similar technique can be used for measuring the length of time between two pulses by configuring the gate to begin counting at the first pulse and to end counting at the second.

Because a 16-bit counter overflows at 65,535 counts, the maximum pulse width measurable with a 1-MHz clock is 65.535 ms. A longer pulse will cause the counter to overflow, producing an inaccurate measurement. A slower clock can be used for longer pulses.

Product Selection Guide

1-MHz PC-based Data Acquisition

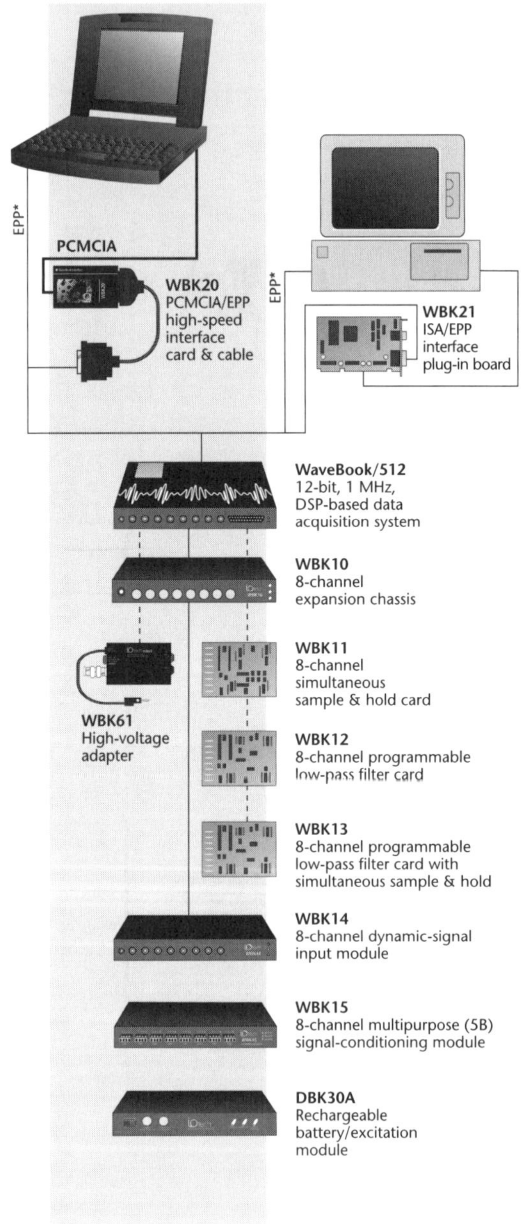

WaveBook/512™

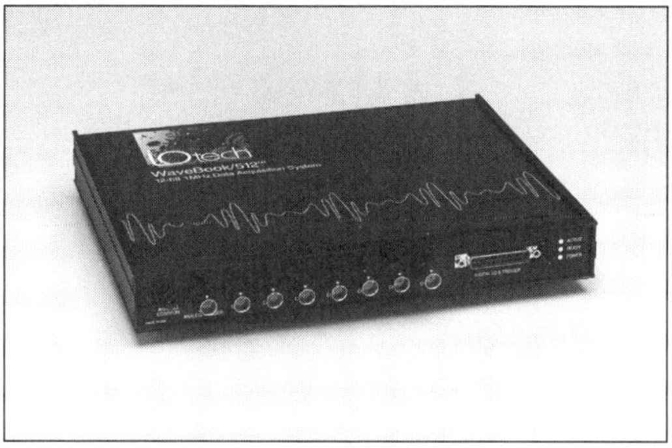

1-MHz, DSP-based Data Acquisition System

- 12-bit 1M-sample/s waveform sampling
- 8 differential inputs, expandable to 72
- 1-μs channel scanning of any combination of channels
- 1-MHz streaming to RAM
- 700-kHz continuous streaming to disk
- 128-location programmable channel/range sequencer
- DSP-based design provides real-time digital calibration on all channels
- Single & multichannel analog triggering with programmable level & slope
- Digital TTL-level triggering
- Programmable pre- & post-trigger sampling rates
- Eight 1-MHz digital inputs
- Connects to notebook PCs for portable operation via enhanced parallel port (EPP) or optional WBK20 PCMCIA card
- Connects to desktop PCs via EPP or optional WBK21 ISA-bus plug-in board
- Accepts one WBK11, WBK12, or WBK13 low-pass filter & simultaneous sample & hold option
- Operable from AC line, a 10 to 30 VDC source, such as a car battery, or optional compact rechargeable battery module

WBK10™

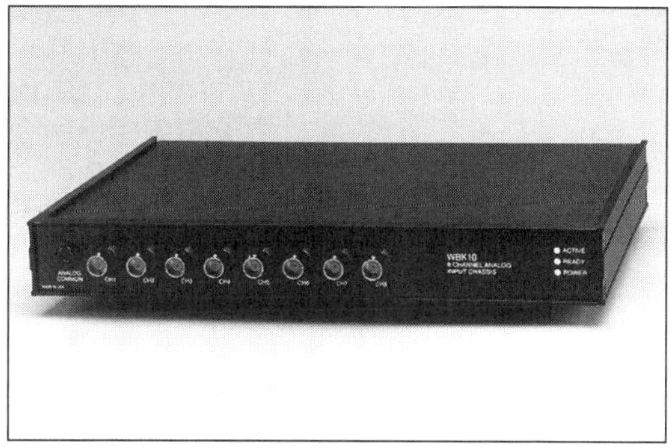

Analog Expansion Module

- Provides seamless expansion of the WaveBook/512 data acquisition system
- Adds 8 differential input channels to the WaveBook/512
- Programmable through the WaveBook/512
- Accepts one WBK11, WBK12, or WBK13 low-pass filter & simultaneous sample & hold option
- Operable from AC line, a 10 to 30 VDC source, such as a car battery, or optional compact rechargeable battery module

WBK11™

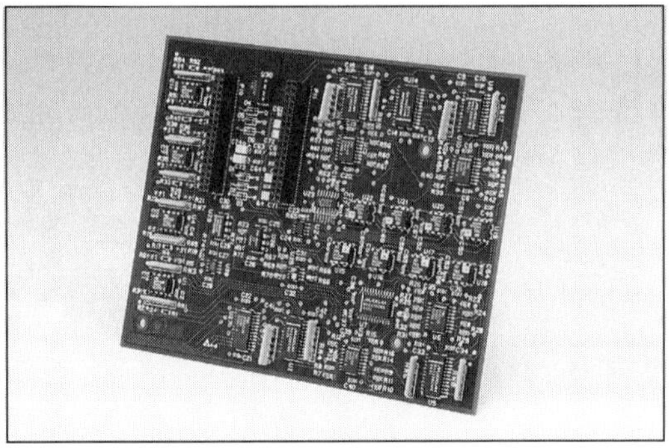

Simultaneous Sample & Hold Card

- Provides simultaneous sample & hold of up to 72 channels for the WaveBook/512 data acquisition system and WBK10 analog expansion module
- Expands the WaveBook/512's gain sensitivity
- Samples all channels within 100 ns of one another
- Provides ranges of:
  - Unipolar: +0.5V, +0.2V, and +0.1V
  - Bipolar: ±0.25V, ±0.1V, ±0.05V

WBK12™, & WBK13™

Programmable Low-Pass Filter Cards

- Provide 8 channels of programmable low-pass anti-aliasing filtering for for the WaveBook/512 data acquisition system and WBK10 analog expansion module
- Provide software programmable 8-pole filters, configurable as elliptic or linear phase
- Provide programmable cutoff frequencies from 400 Hz to 100 kHz
- Accommodate two cutoff frequencies per card, configurable in two 4-channel banks
- Feature full-scale programmable input ranges
  - Unipolar: +100 mV to +10V
  - Bipolar: ±50 mV to ±5V
- Provides 8 input channels with simultaneous sampling (WBK13 only)
- Samples all system channels within 100 ns of one another (WBK13 only)

WBK14™

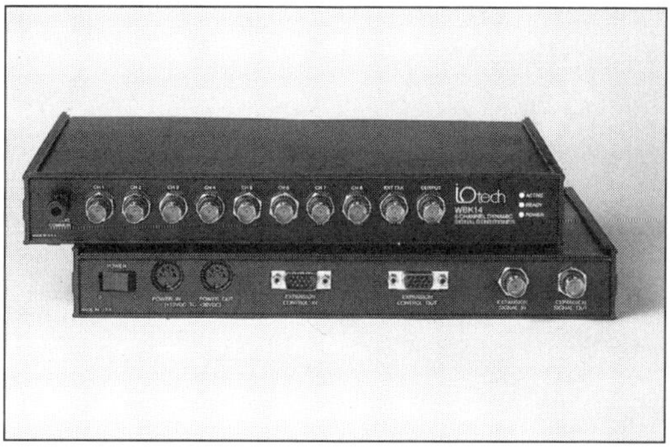

Dynamic Signal Input Module

- Provides eight channels of dynamic signal input for the WaveBook/512 data acquisition system and WBK10 analog expansion module
- Up to 8 WBK14s can be attached to one system
- Offers 8 programmable input ranges, from ±25 mV to ±5V full scale
- Features built-in current source for ICP® biasing
- Includes per-channel simultaneous sample & hold
- Features per-channel, programmable, 8-pole, anti-aliasing low-pass filters with software selectable cut off frequencies from 30 Hz to 100 kHz
- Selectable 0.1 Hz or 10 Hz high-pass filter
- Provides built-in AC excitation source for system calibration & verification
- Operable from AC line, a 10 to 30 VDC source, such as a car battery, or optional compact rechargeable battery module

WBK15™

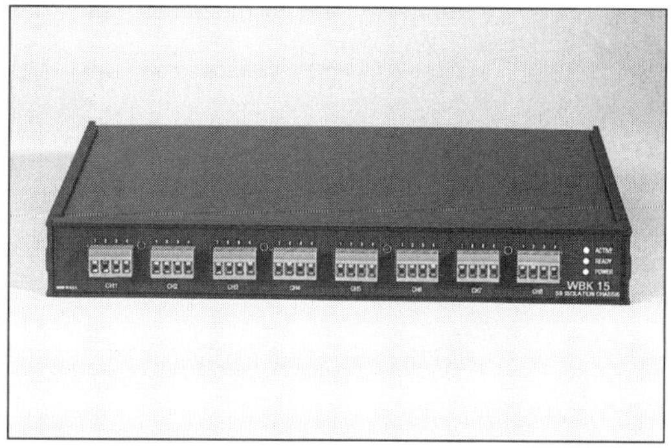

Multi-Purpose Isolated Signal Conditioning Module

- Provides eight 5B isolated-input signal conditioning modules for the WaveBook/512 data acquisition system and WBK10 analog expansion module
- Offers plug-in screw terminal blocks for convenient connection of 5B modules for temperature, strain, acceleration, and a variety of other measurements
- Features on-board cold junction sensing for thermocouple calibration
- Features rugged all-metal construction for portable applications
- Operable from AC line, a 10 to 30 VDC source, such as a car battery, or optional compact rechargeable battery module

WBK61™

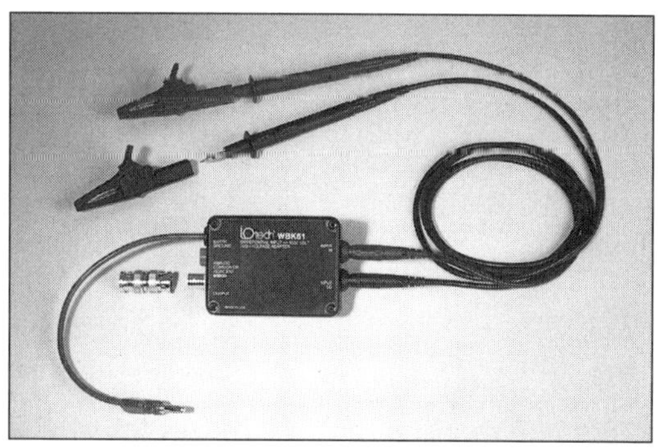

High-Voltage Adapter with Probes

- Provides up to 2000V inputs for the WaveBook/512 data acquisition system and WBK10 analog expansion module
- Includes leads & probes
- Accepts differential inputs

100-kHz PC-based Data Acquisition

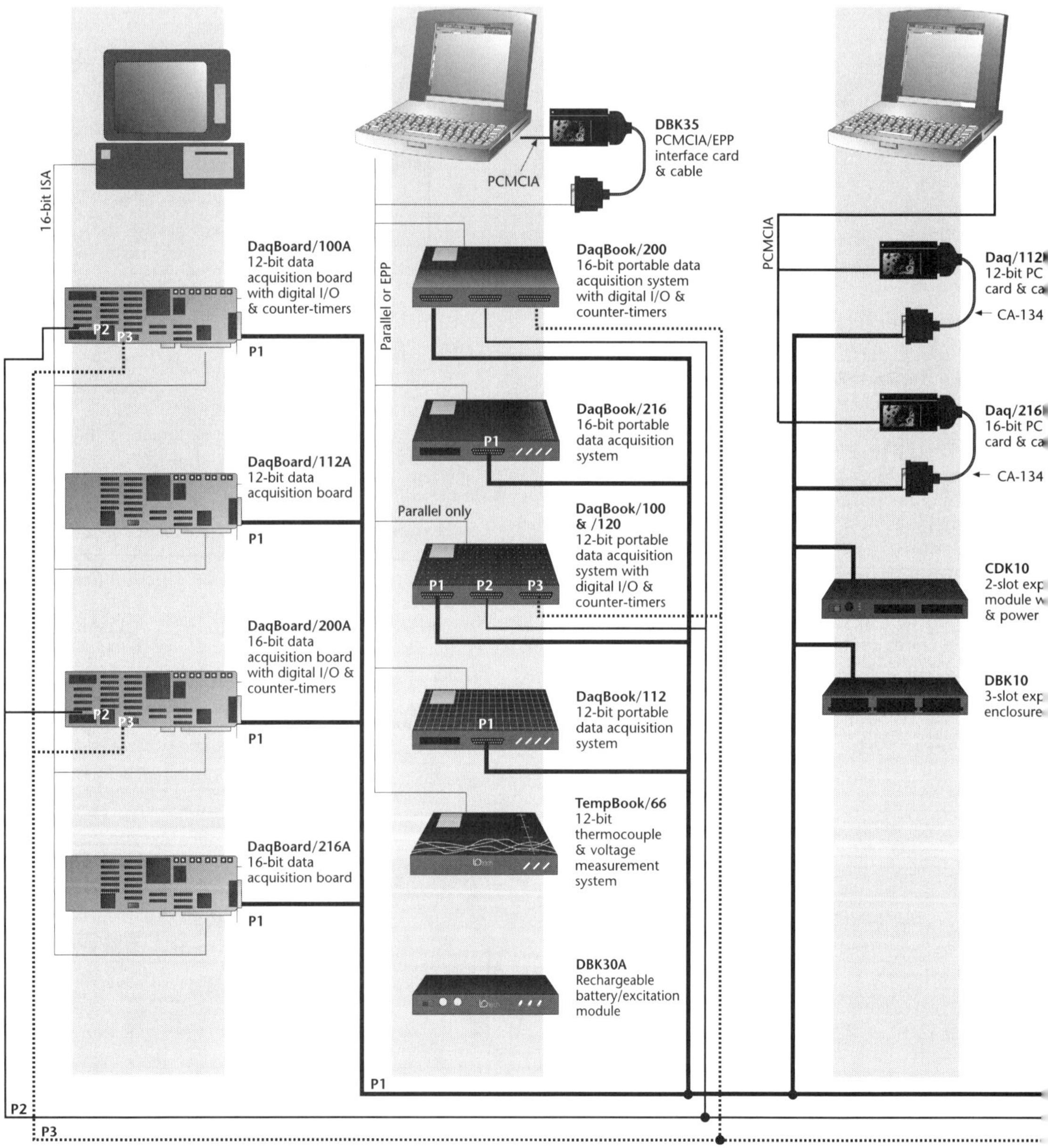

Signal Conditioning & Expansion Options

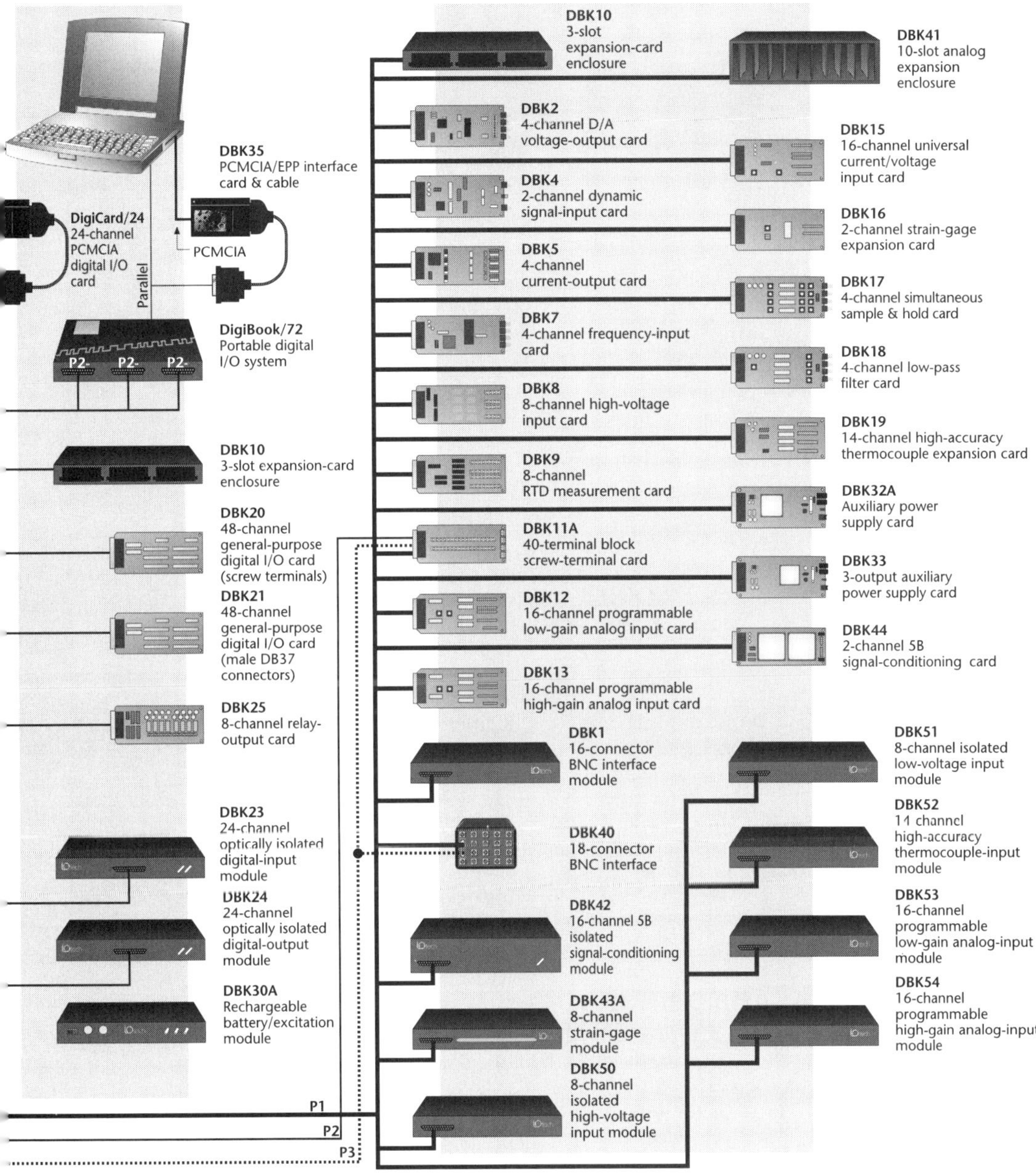

DaqBoard™ Series

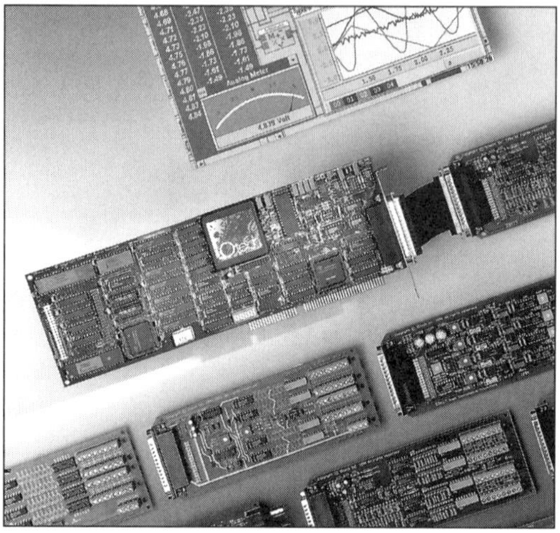

Complete ISA-Bus Based Data Acquisition Solution

- Analog input:
  - 100 kHz A/D converter, 12-bit (100A series) or 16-bit (200A series)
  - 100K reading/s sampling & real-time storage-to-disk
  - 8 differential- or 16 single-ended inputs, expandable to 256 channels
  - channel/gain sequencing at 10 µs intervals, including up to 256 expansion channels
  - 512-location scan memory for user-defined channel/gain sequencing of up to 256 channels
  - x1, 2, 4, or 8 programmable gain (other gains available with option cards)
  - triggerable from analog, digital, or software, including pre-trigger
- Two 12-bit analog outputs with 500 kHz update rate per channel
- Digital I/O:
  - 24 general-purpose digital I/O lines, expandable to 192
  - 16 high-speed digital-input lines, scannable at up to 100 kHz
- 5 programmable 16-bit counter-timers
- Compatible with IOtech's DaqBook & Daq PCMCIA card portable data acquisition systems

DaqBook® Series

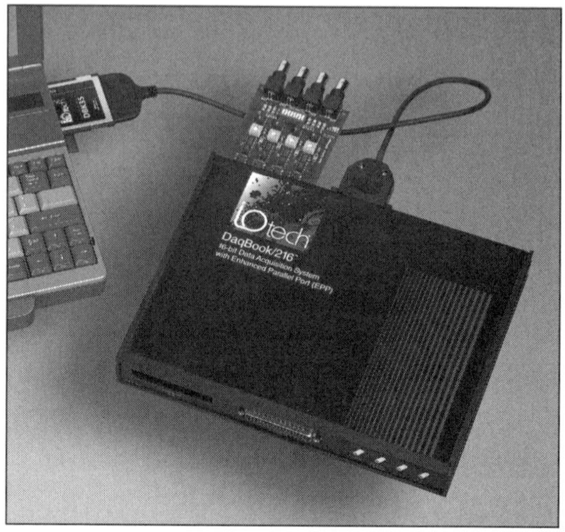

Complete Parallel-Port Based Portable Data Acquisition Solution

- Operable from included AC adapter, optional nickel-cadmium power module, 12V car battery, or any +10 to +20 VDC source
- Analog input:
  - 100 kHz A/D converter, 12-bit or 16-bit
  - 100K reading/s sampling & real-time storage-to-disk
  - 8 differential or 16 single-ended inputs, expandable to 256 channels
  - channel/gain sequencing at 10 µs intervals
  - x1, 2, 4, or 8 programmable gain (other gains available with option cards)
  - 512-location scan memory for user-defined channel/gain sequencing
  - triggerable from analog, digital, or software, including pre-trigger
- Two 12-bit analog outputs
- Digital I/O:
  - 24 general-purpose digital I/O lines, expandable to 192
  - 16 high-speed digital-input lines, scannable at up to 100 kHz
- Five programmable 16-bit counter-timers
- Optional PCMCIA interface
- Compatible with IOtech's DaqBoard and Daq PCMCIA card data acquisition systems for PCs

Daq™ PCMCIA A/D Series

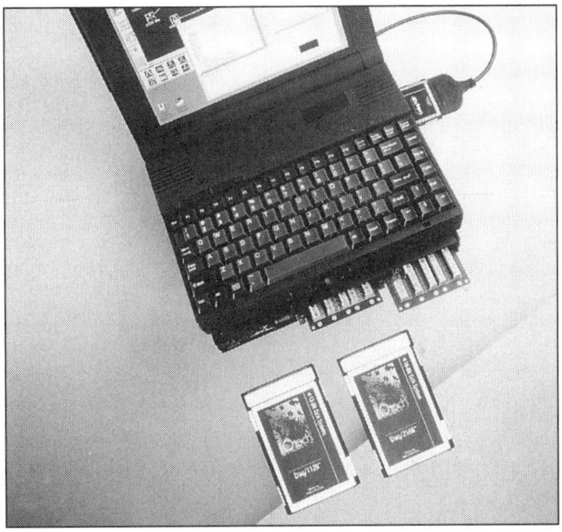

Complete PCMCIA-Based Data Acquisition Solution

- Complies with PC Card Standard Specification 2.1, PCMCIA Type II (5 mm) compatible
- Analog input:
  - 100 kHz A/D converter, 12-bit (Daq/112B) or 16-bit (Daq/216B)
  - 100K reading/s sampling & real-time storage-to-disk
  - 8 diff. or 16 SE inputs, expandable to 256 channels
  - 512-location scan sequencer for user-defined channel/gain sequencing of up to 256 input channels at 10 µs/channel
  - x1, 2, 4, or 8 programmable gain
- 4 digital inputs & 4 digital outputs
- Compatible with IOtech's parallel port-based & ISA-bus -based data acquisition systems

CDK10™

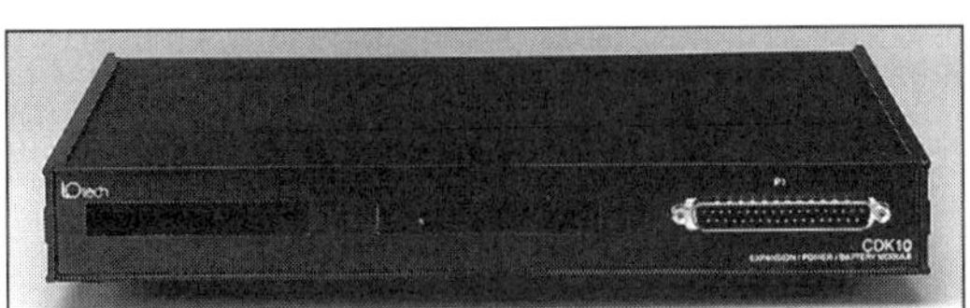

Expansion Chassis for the Daq PCMCIA Cards

- Includes a rechargeable nickel-cadmium battery supply to power expansion cards
- Accepts a wide array of expansion & signal conditioning options

TempBook/66

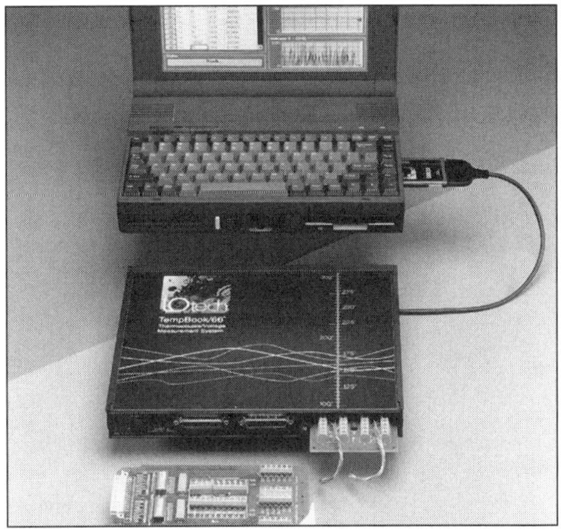

Complete Parallel-Port Based Temperature Measurement Solution

- Provides 8 differential or 16 single-ended voltage inputs
- Features on-board cold-junction & offset-drift compensation
- Supports type J, K, S, T, E, B, R, & N thermocouples
- Operable from included AC adapter, optional nickel-cadmium power module, 12V car battery, or any +10 to +20 VDC source
- Analog input:
  - 12-bit, 100 kHz A/D converter
  - 100K reading/s sampling & real time storage-to-disk
  - 8 differential- or 16 single-ended inputs
  - channel/gain & bipolar/unipolar sequencing at 10 µs intervals
  - 512-location scan memory for user-defined channel/gain sequencing
  - x1, 2, 5, 10, 20, 50, 100, & 200 programmable gains
- Digital I/O:
  - 8 digital input lines, scannable at up to 100 kHz
  - 8 digital output lines
  - 1 trigger input
- One programmable counter-timer
- Optional PCMCIA interface

DigiBook/72

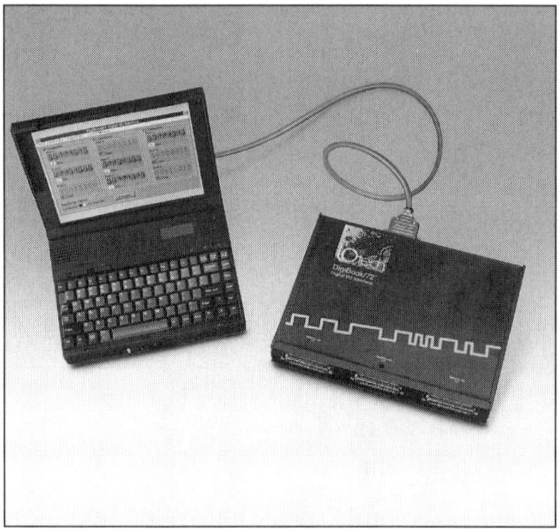

Complete Portable Digital I/O System

- Attaches directly to a PC's parallel port
- Digital I/O capabilities include:
  - 72 programmable TTL-level I/O lines, expandable to 576
- Optically isolated input or output options
- Relay closure output option
- Operable from included AC adapter, optional power module, 12V car battery, or any +10 to +20 VDC source

DigiCard/24

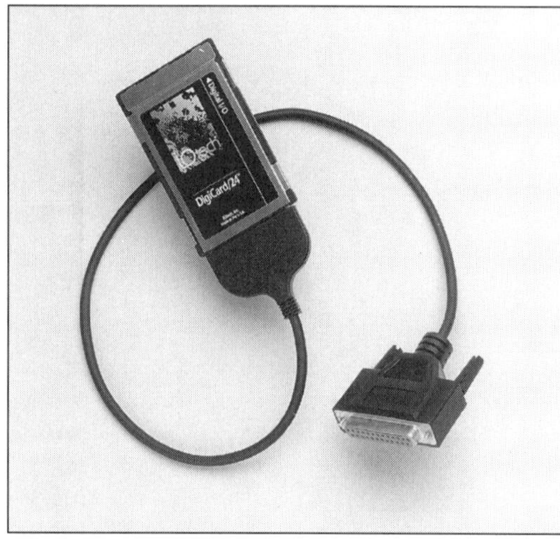

Portable Digital I/O Card

- 24 Digital I/O lines
- PCMCIA Type II compliant

DBK1™

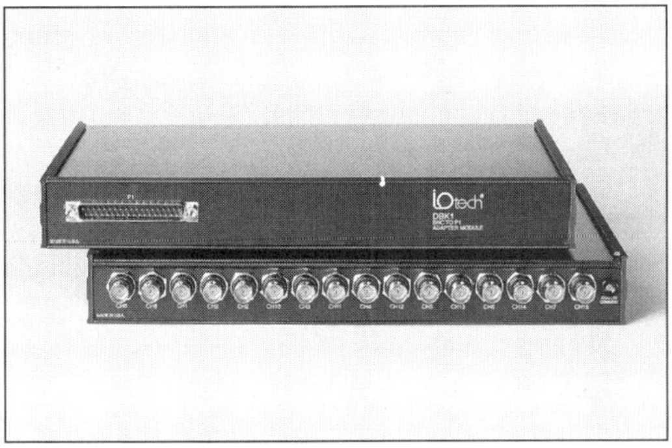

BNC Interface Module

- Provides access to 16 analog inputs via BNC connectors
- Features per-channel 100K Ohm bias resistors for different inputs

DBK4™

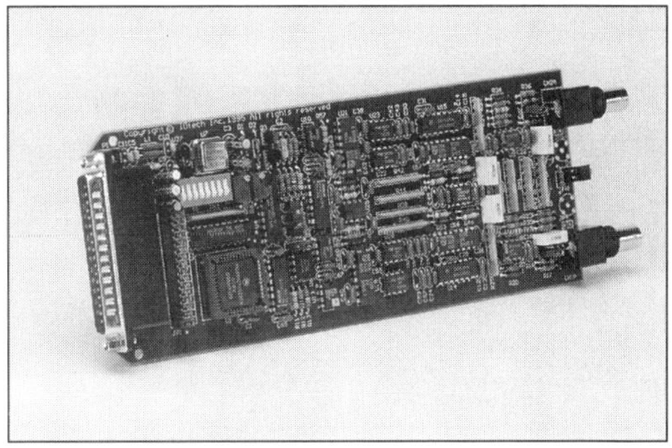

Dynamic Signal-Input Card

- Provides 2 dynamic analog inputs/per-channel SS&H
- Offers per-channel PGA for ±50 mV, ±500 mV, or ±5V FS input
- Includes a built-in current source for ICP transducer biasing
- Offers per-channel anti-aliasing filters with software programmable cutoff frequencies from 141.6 Hz to 18 kHz
- Accommodates AC or DC coupled signals
- Dynamic analog inputs sampled within 50 ns of each other

DBK2™

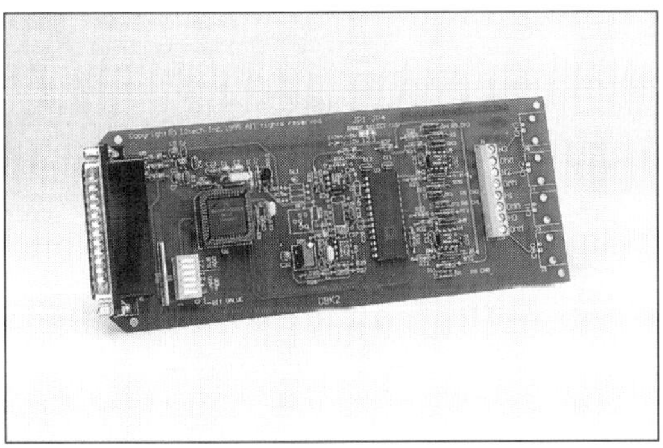

Voltage-Output Card

- 14-bit, quad analog output
- Features on-board bipolar reference for accurate & stable voltage outputs
- Provides per-channel bipolar voltage ranges

DBK5™

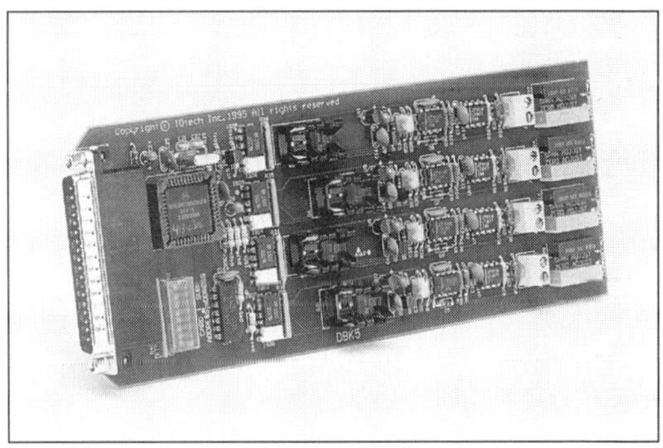

Current-Output Card

- Provides four isolated 4 to 20 mA current outputs
- Complies with current loop resistances of up to 1950 Ohms
- Compatible with 12 to 45V current loop supplies
- Features predictable power-on & fault states
- Provides 12-bit resolution

DBK7™

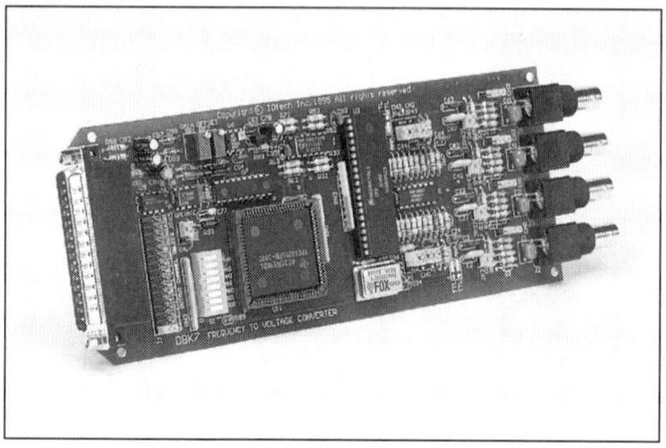

Frequency-Input Card

- Provides four frequency-measurement channels
- Programmable from 1 Hz to 950 kHz per channel
- Provides frequency resolution to 0.00025 Hz
- Accommodates low-level, high-level, or digital inputs
- Provides per-channel, user-configurable low-pass filters

DBK8™

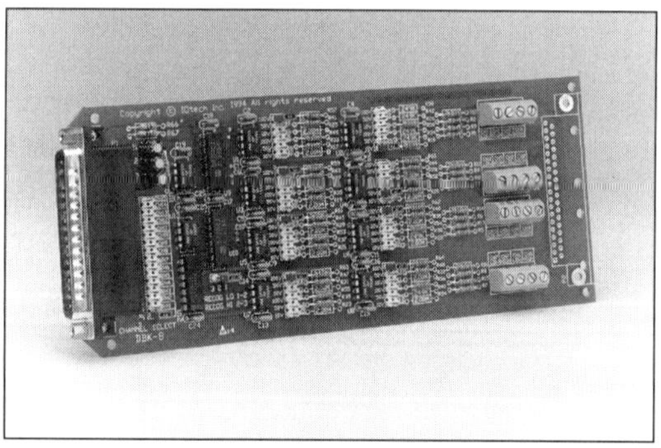

High-Voltage Input Card

- Provides eight high-voltage input channels
- Per-channel selection of three voltage ranges via jumper
- Provides 10M Ohm input impedance for minimum loading errors
- Buffered inputs on each channel for high-speed scanning

DBK9™

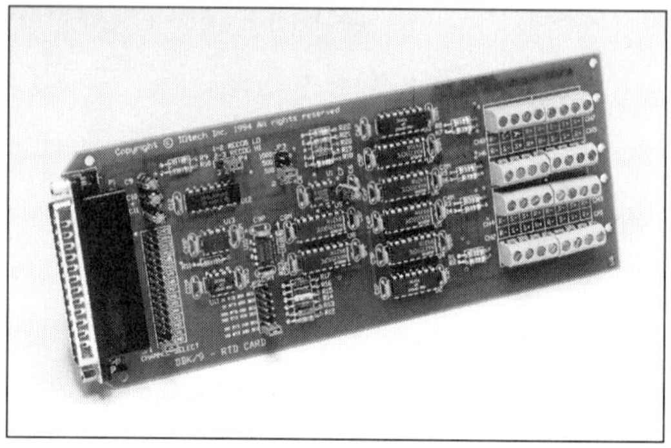

RTD Measurement Card

- Provides eight RTD channels
- Supports 3- & 4-wire RTDs
- Supports three ranges of RTD resistance

DBK10™

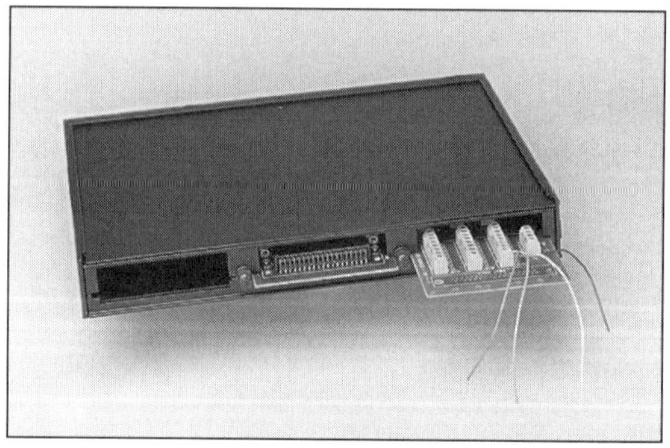

Expansion Card Enclosure

- Accommodates any three analog or digital DBK cards
- Ideal for applications that require six or fewer DBK cards

DBK11A™

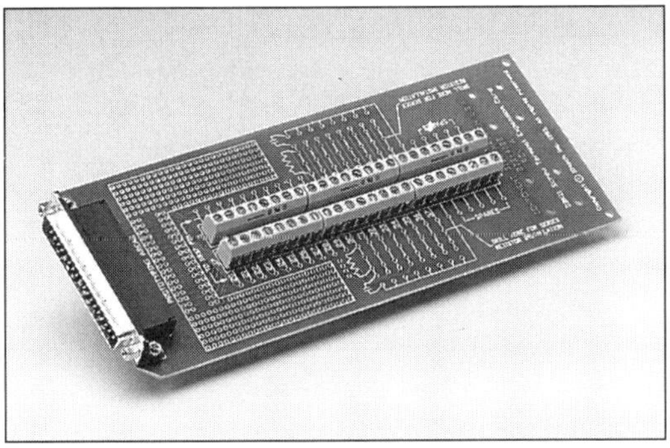

Screw-Terminal Card

- Provides screw-terminal connection for application inputs
- Provides four user-installable BNC connectors

DBK12 & DBK13™

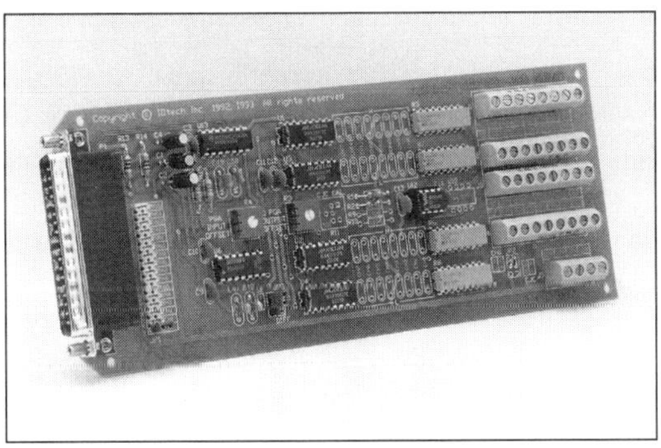

Analog Input Cards

- Provide 16 differential-ended inputs
- DBK12 offers x1, 2, 4, or 8 programmable gain
- DBK13 offers x1, 10, 100, or 1000 programmable gain

DBK15™

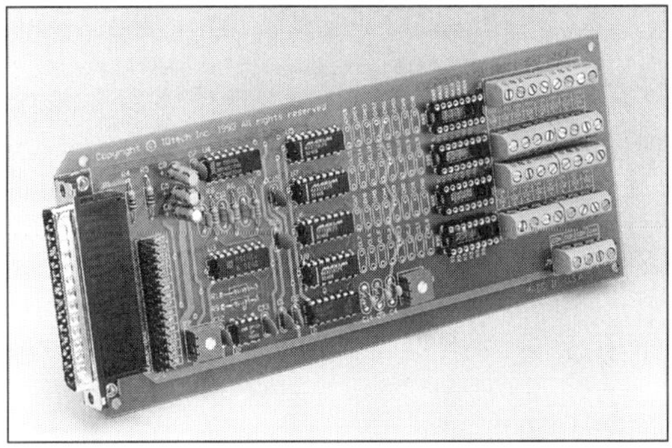

Universal Current/Voltage Input Card

- Ideally suited for 4 to 20 mA measurements
- Can measure up to ±30V*
- Provides 16 differential-ended input channels
- Offers user-configurable current or voltage input

* ±10V to 30V input signals can be measured by installing <10K Ohms attenuation resistors on the card. To accurately measure high-voltage signals with >100 Ohms of output impedance, the DBK8 high-voltage input card should be used rather than the DBK15.

DBK16™

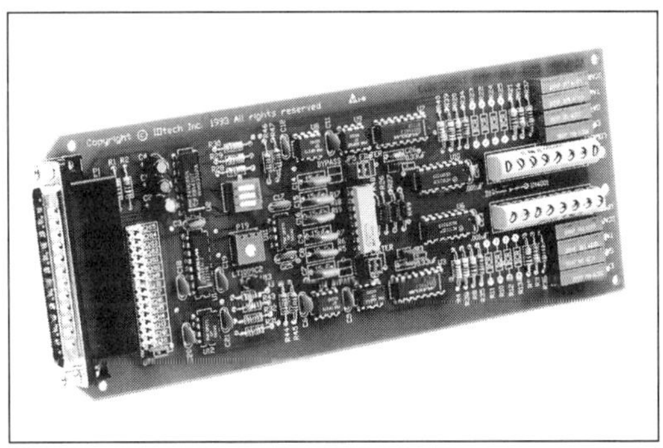

Strain-Gage Expansion Card

- Provides 2 channels of strain-gage inputs
- Accommodates most bridge type sensors, including 4-element full bridges

DBK17™

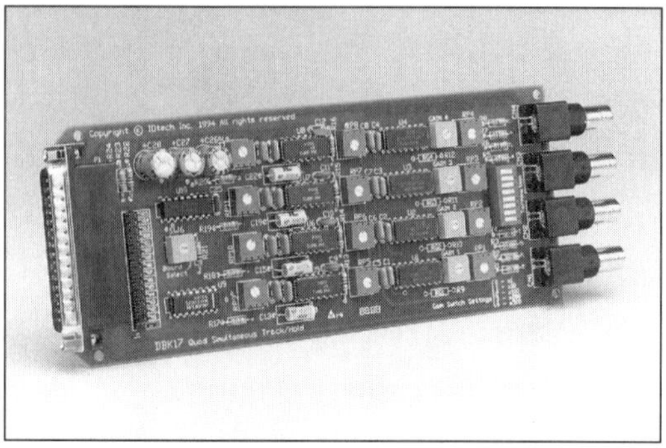

Simultaneous Sample & Hold Card

- Provides four input channels with simultaneous sample & hold
- Features a separate instrumentation amplifier & input stage for each channel

DBK19™

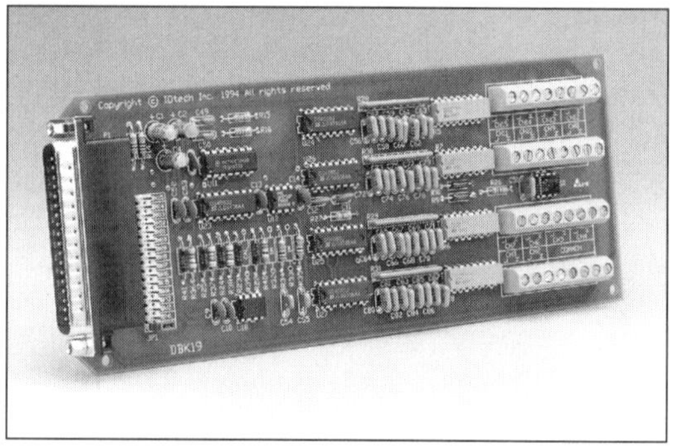

Thermocouple Expansion Card

- Provides 14 thermocouple inputs
- Features on-board cold-junction & offset-drift compensation
- Supports type J, K, S, T, E, B, R, & N thermocouples

DBK18™

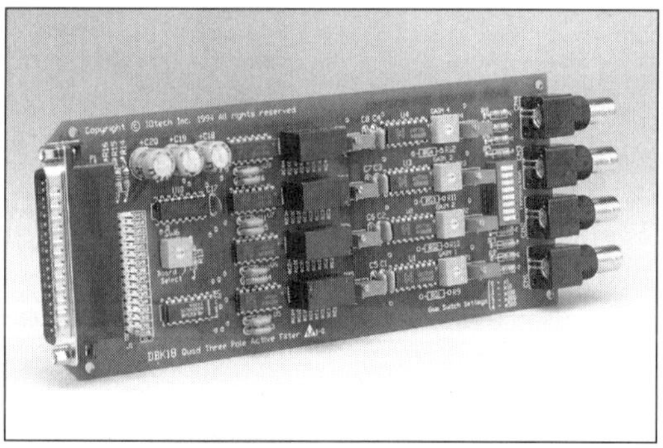

Low-Pass Filter Card

- Provides four independent three-pole low-pass filter channels
- Provides a separate instrumentation amplifier input stage for each channel
- User-configurable from DC to 50 kHz cut-off

DBK20™ & DBK21™

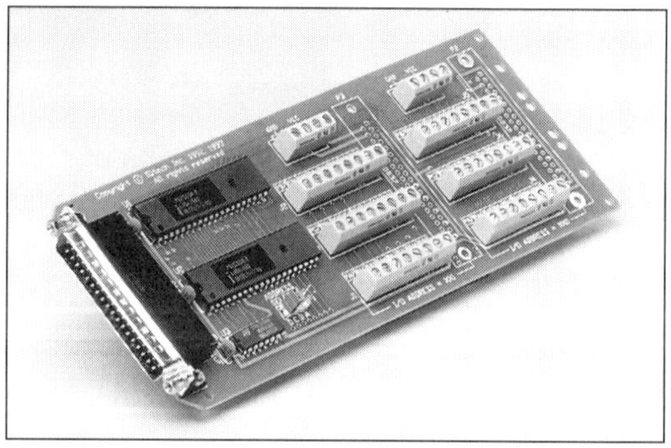

Digital I/O Cards

- Offers 48 I/O lines expandable in 8-bit groups as inputs or outputs
- Channels expandable in 8-bit groups as inputs or outputs
  - DBK20 features screw-terminal connectors
  - DBK21 features male DB37 connectors

DBK23™

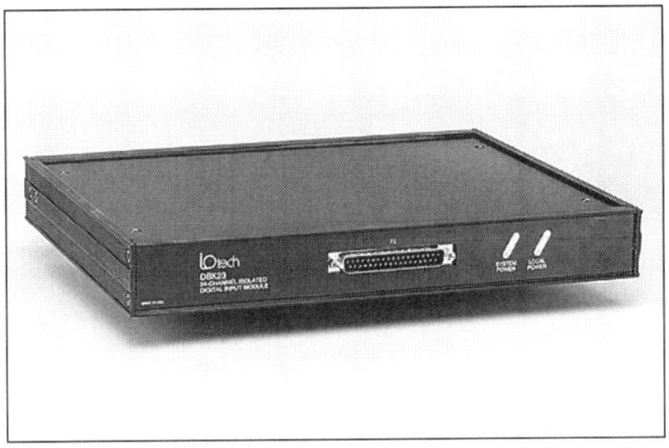

Isolated Digital Input Module

- Provides twenty-four isolated digital inputs
- Optically isolated inputs accept up to 30 VDC
- 500V channel-to-channel isolation
- 500V channel-to-system isolation
- Expandable to 192 digital lines

DBK25™

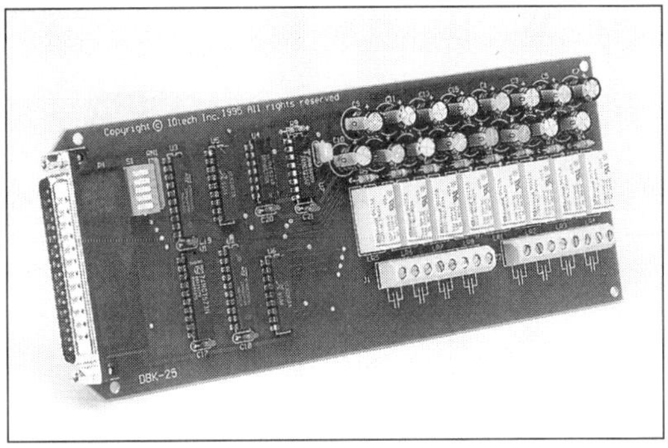

Relay-Output Card

- Provides eight independent relay contact outputs
- Requires minimal system power

DBK24™

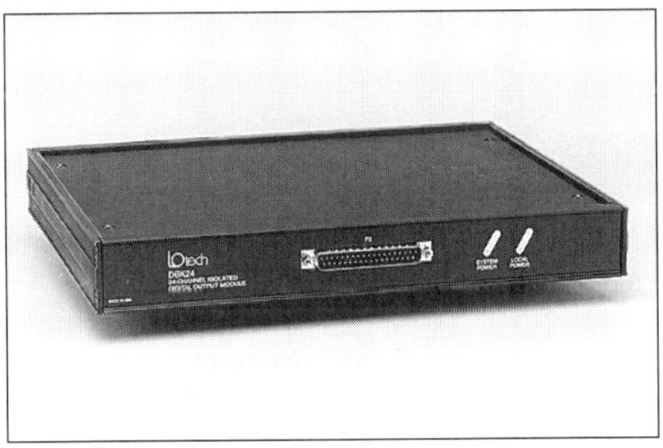

Isolated Digital-Output Module

- Provides twenty-four isolated digital outputs
- Optically isolated outputs up to 60 VDC
- 500V channel-to-channel isolation
- 500V channel-to-system isolation
- Expandable to 192 digital lines

DBK41™

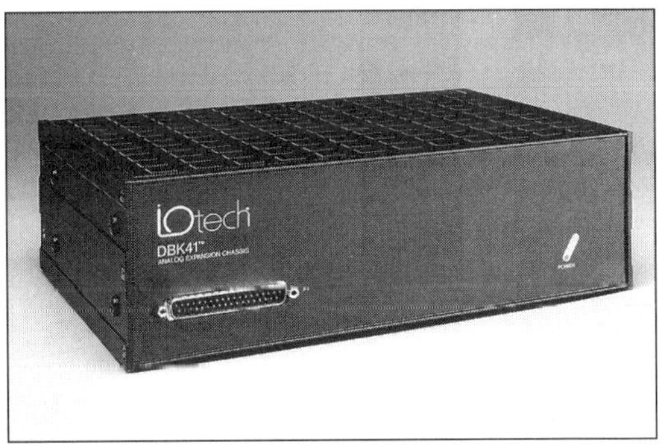

Expansion Chassis

- Accepts up to ten DBK analog expansion cards
- Features analog backplane, eliminating the need for a cascaded ribbon cable
- Enables easy card installation
- Features rugged all metal construction for portable applications
- Attaches to the DaqBoard & DaqBook P1 expansion connector

DBK42™

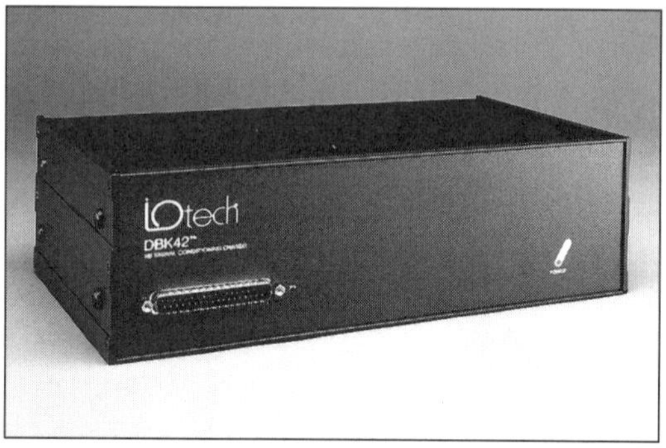

Isolated Signal Conditioning Module

- Accepts any combination of up to sixteen 5B isolated input signal conditioning modules
- Attaches to the DaqBoard, DaqBook, or Daq PCMCIA card P1 expansion connector
- Optional screw-terminal connection block includes cold-junction sensors for thermocouple applications
- Features rugged all-metal construction for portable applications
- Includes a built-in power supply, operable from 10 to 24 VDC

DBK44™

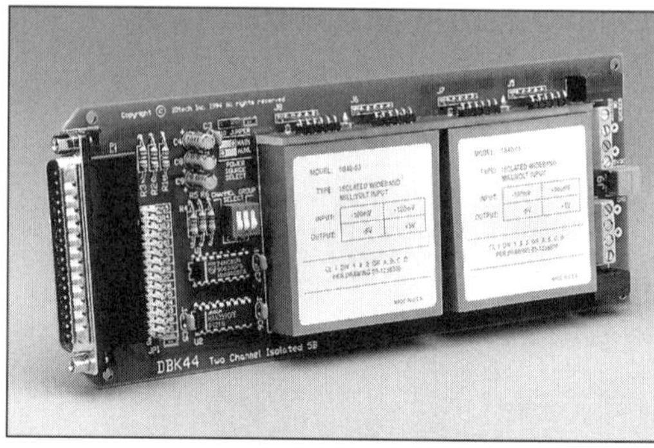

Isolated Signal Conditioning Card

- Accepts two 5B isolated input signal conditioning modules of any type
- Features on-board screw-terminal signal connections, plus cold-junction sensors for thermocouple applications
- Installs into a DBK10 three-slot or DBK41 ten-slot expansion-card enclosure, or a DaqBook/112 or a DaqBook/216 expansion slot

DBK43A™

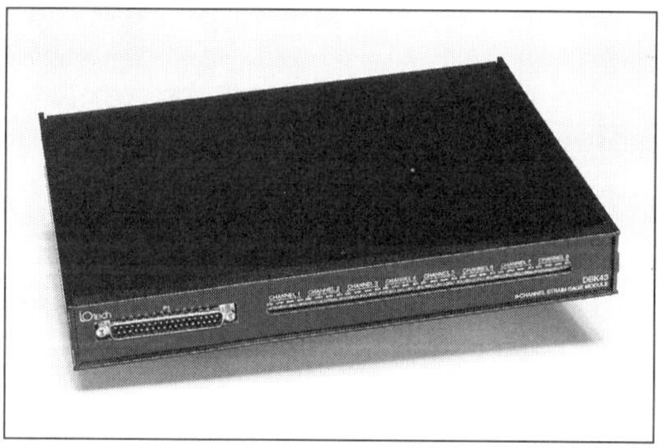

Strain-Gage Module

- Provides eight channels of strain-gage input
- Accommodates most bridge-type sensors, including 4-element full bridges
- On-board shunt-calibration provisions enhance setup of high-channel count systems

DBK50™

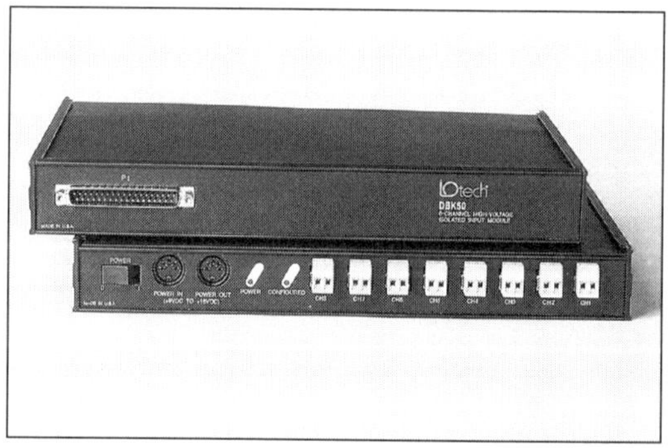

Isolated High-Voltage Input Module

- Provide eight channels of isolated high-voltage input
- Offers bipolar input ranges: 10V, 100V, & 300V full scale input, software programmable
- Provides 750V channel-to-channel isolation and 1250V channel-to-system isolation
- Operable from included AC adapter, optional nickel-cadmium power module, 12V car battery, or any +10 to +20 VDC source

DBK51™

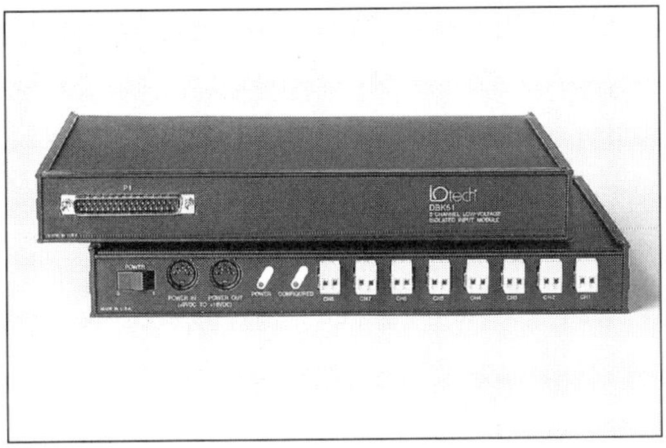

Isolated High-Voltage Input Module

- Provide eight channels of isolated low-voltage input
- Offers bipolar input ranges: 100 mV, 1V, & 10V full scale input, software programmable
- Provides 750V channel-to-channel isolation and 1250V channel-to-system isolation
- Operable from included AC adapter, optional nickel-cadmium power module, 12V car battery, or any +10 to +20 VDC source

DBK52™

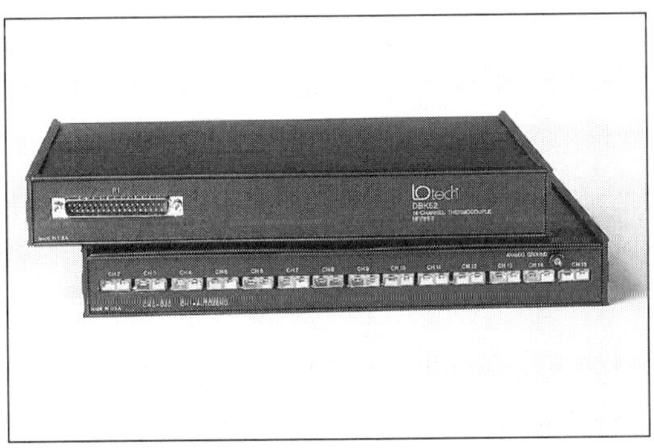

Thermocouple-Input Module

- Provides fourteen thermocouple-inputs via convenient thermocouple connectors
- Features on-board cold-junction & offset-drift compensation
- Supports type J, K, S, T, E, B, R, & N thermocouples

DBK53™

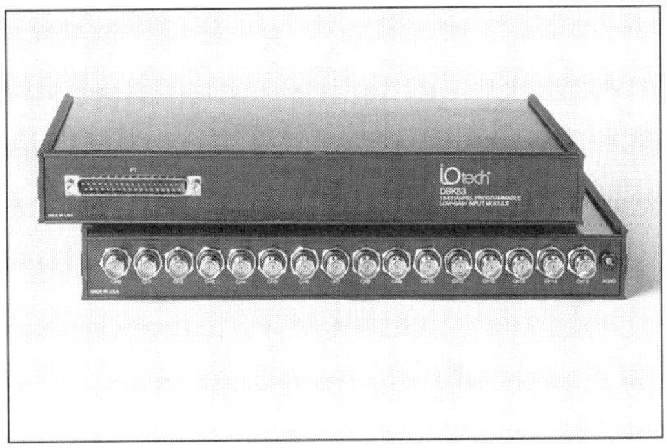

Low-Gain Analog-Input Module

- Provides 16 differential or single-ended inputs
- Programmable gain ranges of x1, x2, x4, & x8
- Equipped with convenient BNC input connectors

DBK54™

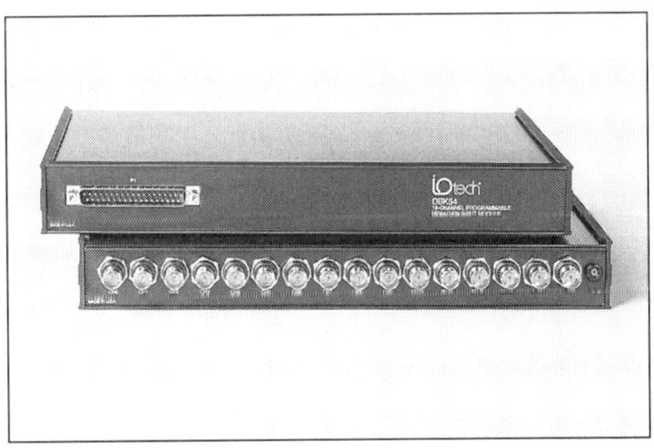

High-Gain Analog-Input Module

- Provides 16 differential- or single-ended inputs
- Programmable gain ranges of x1, x10, x100, & x1000
- Convenient BNC input connectors

High-Speed Temperature & Voltage Instruments

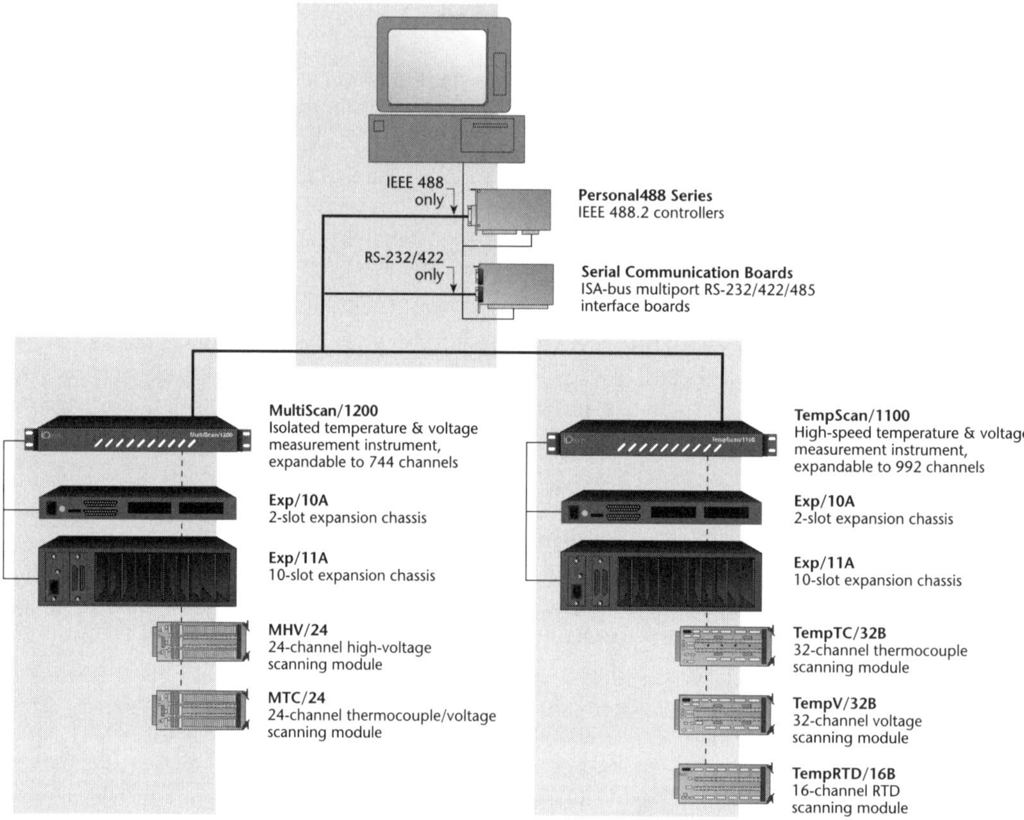

TempScan/1100

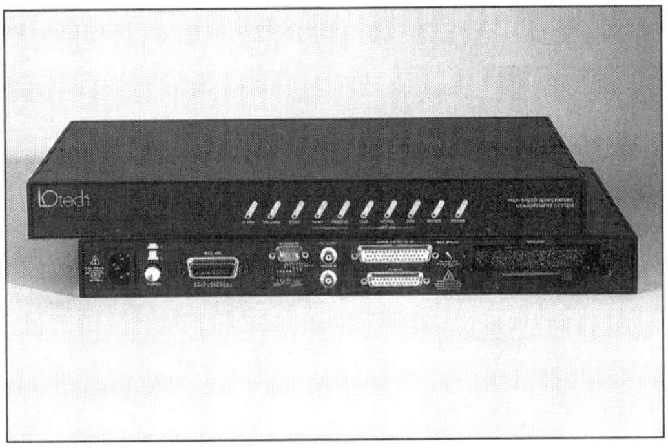

High-Speed Temperature & Voltage Instrument

- Measures thermocouples, volts, & RTDs at up to 960 channels/s
- Accepts optional scanning modules for measuring thermocouples, RTDs or DC volts
- Expandable up to 992 channels
- IEEE 488 & RS-232/422 interfaces
- 32 TTL digital alarm outputs & 8 TTL-compatible digital inputs
- Custom thermocouple types for user-defined linearization tables
- Two programmable scan rates for:
  - pre-trigger & post-trigger sampling
  - accelerated sampling on-event detection
- 256 Kbytes of memory, expandable up to 8 Mbytes
- Built-in real-time clock:
  - synchronizes acquisition to time of day
  - provides time & date stamping for trend monitoring

TempV/32B

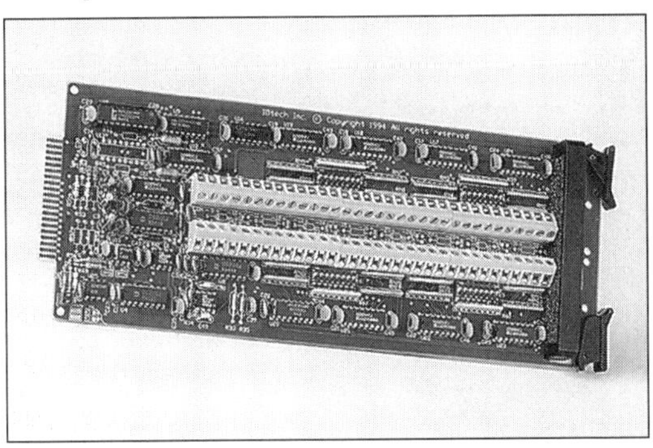

Voltage Scanning Module

- Contains 32 differential input channels
- Capable of measuring voltage with programmable ranges of ±10V, ±5V, ±1V, and ±100 mV

TempTC/32B

Thermocouple Scanning Module.

- Contains 32 differential input channels
- Accommodates J, K, T, E, R, S, B, and N, thermocouple types
- Measurements may be designated in units of °C, °F, °K, °R , or volts.

TempRTD/16B

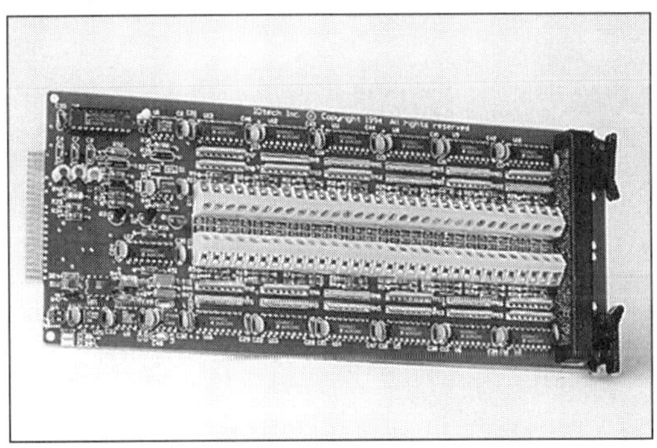

RTD Scanning Module

- Supports 16 channels of 3 or 4-wire RTDs
- Measurements may be designated in units of °C, °F, °K, or °R.

MultiScan/1200™

Isolated Temperature & Voltage Instrument

- Measures isolated temperature, DC volts, AC volts, & waveforms in one compact instrument
- Scans thermocouples & DC volts at up to 147 channels/s
- Single-channel burst mode for digitizing waveforms at rates up to 20 kHz
- Two scanning modules available for measuring 24 channels of thermocouples/volts or high voltage, respectively
- Expandable up to 744 channels
- IEEE 488 & RS-232/422 interfaces
- 32 TTL digital alarm outputs & 8 TTL-compatible digital inputs
- Custom thermocouple types for user-defined linearization tables
- Two programmable scan rates for:
  - pre-trigger & post trigger sampling
  - accelerated sampling on-event detection
- 256 Kbytes of memory, expandable up to 8 Mbytes
- Built-in real-time clock:
  - synchronizes acquisition to time of day
  - provides time & date stamping for trend monitoring

MTC/24

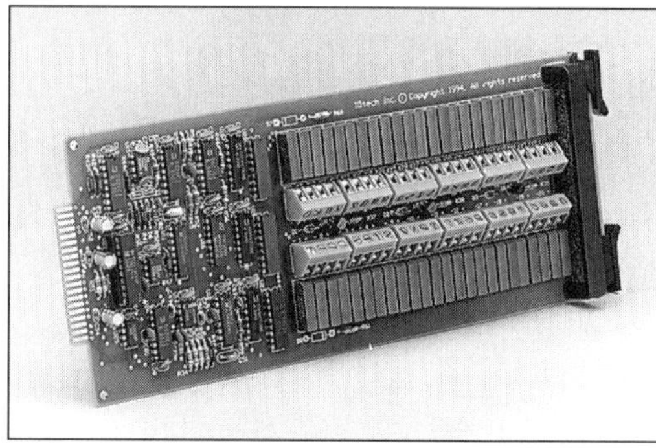

Thermocouple/Volt Scanning Module

- Contains 24 differential input channels
- Accommodates J, K, T, E, R, S, B, and N, thermocouple types or ±10 VFS, ±5 VFS, ±1 VFS, and ±100 mVFS inputs
- Channel-to-channel isolation is 200 VDC peak

MHV/24

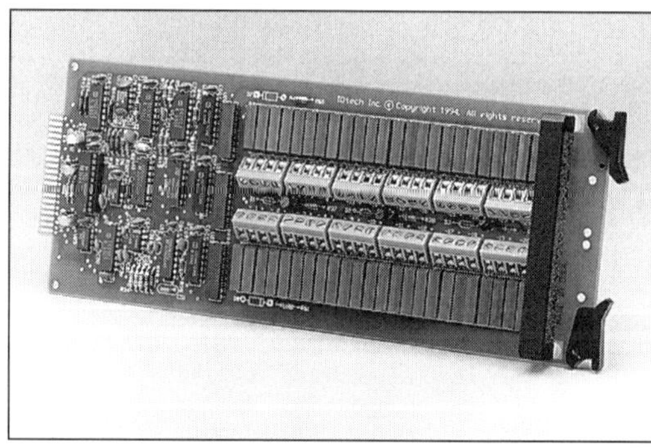

High-Voltage Scanning Module

- Contains 24 differential input channels
- Capable of measuring voltage with programmable ranges of ±250 VFS, ±25 VFS, and ±2.5 VFS
- Channel-to-channel isolation is 500 VDC peak.

WaveView

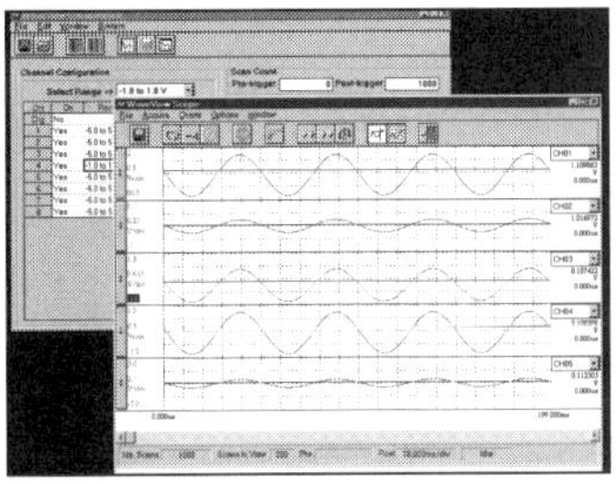

- Windows-based setup and acquisition application for the 1-MHz, DSP-based WaveBook/512 data acquisition system
- Allows you to configure your system, display data and save data to disk

DaqView

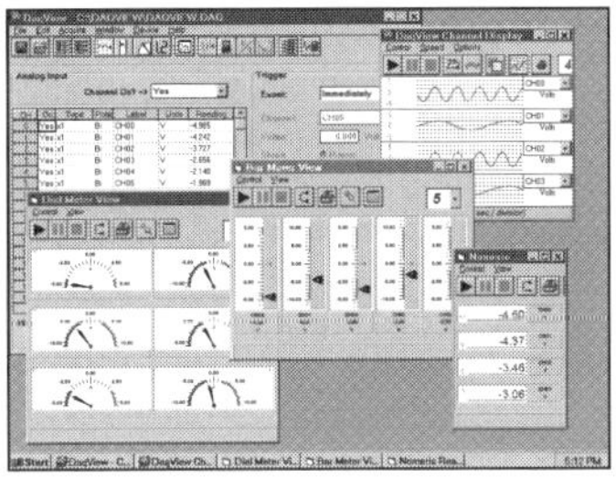

- Windows-based setup and acquisition application for the 100-kHz, ISA-bus based DaqBoard, parallel-port-based DaqBook and Daq PCMCIA data acquisition systems
- Allows you to configure your system, display data and save data to disk

DaqViewXL

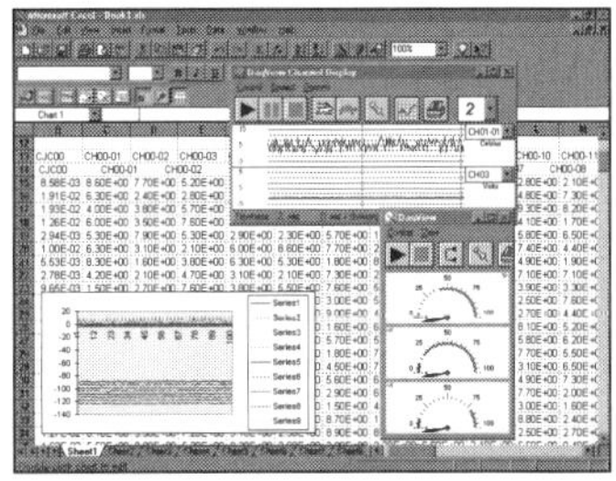

- Excel add-in that provides complete data acquisition functionality for the 100-kHz, ISA-bus based DaqBoard, parallel-port-based DaqBook and Daq PCMCIA data acquisition systems
- Provides data acquisition capability to Excel

TempView

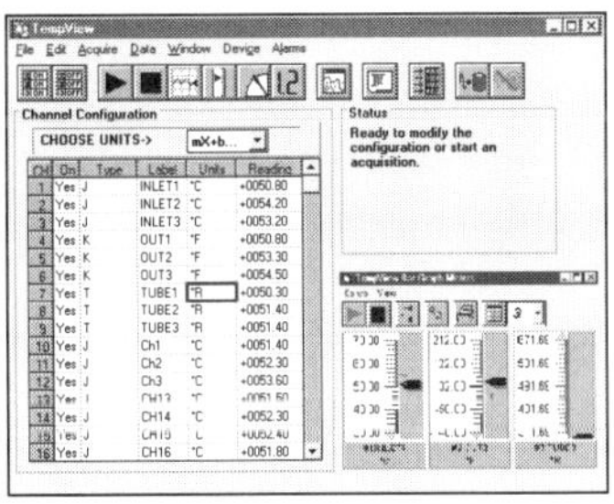

- Windows-based setup and acquisition application for the TempScan/1100 and the MultiScan/1200 high-speed temperature and voltage measurement instruments
- Allows you to configure your system, display data and save data to disk

PostView

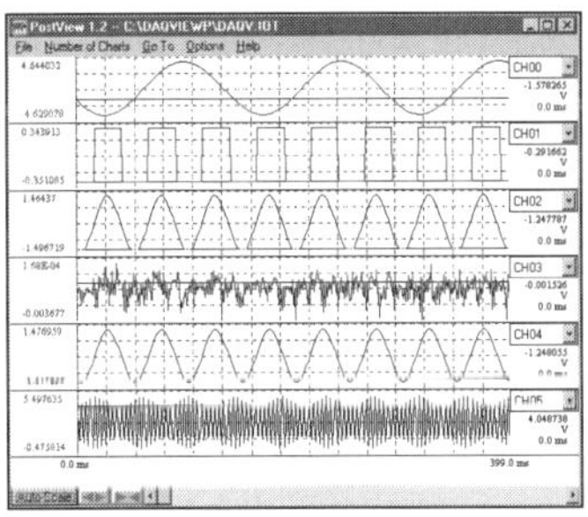

- Windows-based post acquisition data analysis and review software, ships with the WaveBook/512, DaqBoard, DaqBook, Daq PCMCIA card, Tempscan/1100 and MultiScan/1200 systems
- Provides stripchart-recorder-like graphical displays for up to 16 channels of previously acquired data

REFERENCES

Manual on the Use of Thermocouples in Temperature Measurement, ASTM MNL12, 4th ed., American Society for Testing and Materials, Philadelphia, 1993.

Thermocouple Reference Tables based on the IPTS-68, Publication No. NBS MN-125, National Institute of Standards and Technology, 1974.

THE NICROSIL VERSUS NISIL THERMOCOUPLE : Properties and Thermoelectric Reference Data, Publication No. NBS MN-161, National Institute of Standards and Technology, 1978.

THE TEMPERATURE HANDBOOK, VOL 29, Omega Engineering Inc., Stamford, CT., 1995

Trademarks

The following are trademarks of IOtech, Inc.:

CDK10, Daq/112B, Daq/216B, DaqBoard, DaqBoard/100A, DaqBoard/112A, DaqBoard/200A, DaqBoard/216A, DaqBook, DaqBook/100, DaqBook/100/OEM, DaqBook/112, DaqBook/112/OEM, DaqBook/120, DaqBook/120/OEM, DaqBook/200, DaqBook/200/OEM, DaqBook/216, DaqBook/216/OEM, DaqView, DaqViewXL, DBK1, DBK2, DBK4, DBK5, DBK7, DBK8, DBK9, DBK10, DBK11A, DBK12, DBK13, DBK15, DBK16, DBK17, DBK18, DBK19, DBK20, DBK21, DBK23, DBK24, DBK25, DBK30A, DBK32A, DBK33, DBK35, DBK40, DBK41, DBK42, DBK43A, DBK44, DBK50, DBK51, DBK52, DBK54, DigiBook/72, DigiBook/72/OEM, DigiCard/24, DigiView, EPPCard/1, Exp/10A, Exp/11A, LPTCard/1, MHV/24, MTC/24, MultiScan/1200, MultiMEM1, MultiMEM4, MultiMEM8, PostView, TBK10, TempBook/66, TempScan/1100, TempMEM1, TempMEM4, TempMEM8, TempRTD/16A, TempRTD/16B, TempTC/32A, TempTC/32B, TempV/32A, TempV/32B, TempView, WaveBook/512, WaveView, WBK10, WBK11, WBK12, WBK13, WBK14, WBK15, WBK20, WBK21, WBK61

All other trademarks are property of their respective holders.

Specifications are subject to change without notice.

Index

Symbols

A

B

C

D

H

I

K

L

M

N

O

P

Q

R

S

Index

Fold here

NO POSTAGE
NECESSARY IF
MAILED IN THE
UNITED STATES

BUSINESS REPLY MAIL
FIRST-CLASS MAIL PERMIT NO. 4417 CLEVELAND OH

POSTAGE WILL BE PAID BY ADDRESSEE

IOtech®

PO BOX 391345
CLEVELAND OH 44139-9846

Fold here